The Cambridge Manuals of Science and Literature

BREWING

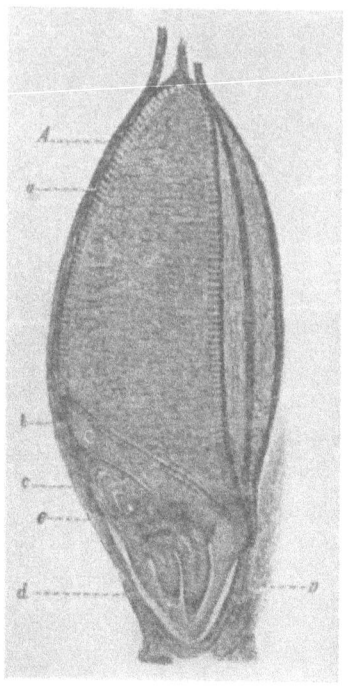

Longitudinal Section of Barleycorn.
(From Malzbereitung und Bierfabrikation.—J. E. Thausing.)

- *A* Husk.
- *B* Endosperm (starch body).
- *a* Aleurone layer.
- *b* Scutellum.
- *c* Plumule (acrospire).
- *C* Germ (embryo).
- *D* Basal bristle.
- *d* Radicle.
- *e* Axis of Germ.

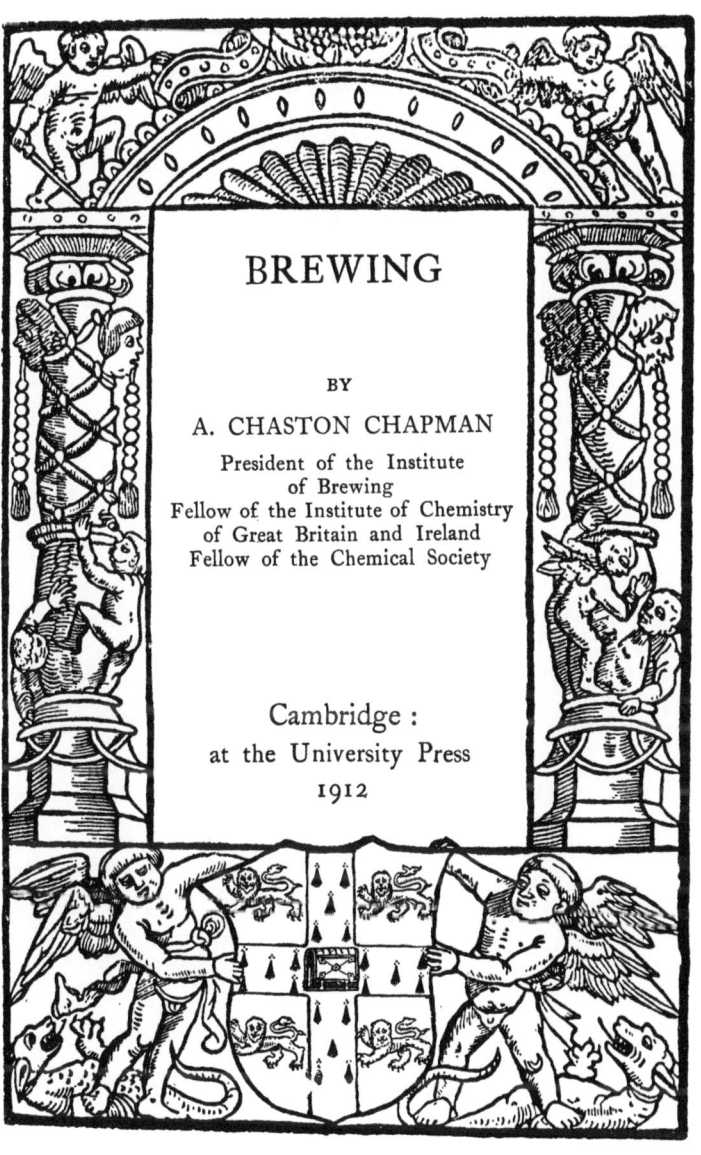

BREWING

BY

A. CHASTON CHAPMAN

President of the Institute
of Brewing
Fellow of the Institute of Chemistry
of Great Britain and Ireland
Fellow of the Chemical Society

Cambridge:
at the University Press
1912

CAMBRIDGE UNIVERSITY PRESS
Cambridge, New York, Melbourne, Madrid, Cape Town,
Singapore, São Paulo, Delhi, Tokyo, Mexico City

Cambridge University Press
The Edinburgh Building, Cambridge CB2 8RU, UK

Published in the United States of America by Cambridge University Press, New York

www.cambridge.org
Information on this title: www.cambridge.org/9781107605954

© Cambridge University Press 1912

First published 1912
First paperback edition 2011

A catalogue record for this publication is available from the British Library

ISBN 978-0-521-04617-6 Hardback
ISBN 978-1-107-60595-4 Paperback

Cambridge University Press has no responsibility for the persistence or accuracy of URLs for external or third-party internet websites referred to in this publication, and does not guarantee that any content on such websites is, or will remain, accurate or appropriate.

With the exception of the coat of arms at the foot, the design on the title page is a reproduction of one used by the earliest known Cambridge printer, John Siberch, 1521

PREFACE

GREAT as is the debt of gratitude which the brewing industry owes to the labours of scientific men, it has been more than repaid by the immense services which that industry has indirectly rendered to the advancement of modern science. It may be said without exaggeration that in respect of the number of scientific investigations of the first order of importance to which it has given rise, the brewing industry stands easily pre-eminent among the industries of mankind, and that without the stimulus furnished by the desire to arrive at the meaning of some of the more important phenomena connected with the brewing of beer, both chemical and biological science would probably be the poorer to-day by some of their most valued intellectual achievements. In support of this statement it is only necessary to refer to the investigation of the chemical and other changes occurring during the growth of the barleycorn, the elucidation of the mechanism of the hydrolysis of starch by diastase, the nature of enzyme action, and last, but not least, the numerous researches into the nature of fermentation, with all the wonderful chemical and biological facts which those researches have revealed. Whilst all this is indisputably true, it is unfortunately a fact that about no other industry do so much ignorance and misconception exist, even on the part of intelligent and well-educated people. The brewing of beer is regarded by many as an operation of a simple and more or less mechanical description, which is not of sufficient importance to merit study or of sufficient interest to claim a share of their attention. It is in the hope of doing something, even though it be but little, to correct this widely-spread impression, that I have most willingly accepted the invitation to contribute this little work to *The Cambridge Manuals of Science and Literature*. To prevent any misunderstanding, I may say at once that it is not intended to serve as a text-book for the student of brewing and still less as a hand-book for the practical brewer. In it I have only referred to methods of practical working when such reference was necessary for the purpose of affording a general view of the process, or appeared to render the application of certain principles more intelligible. I have in fact confined myself as far as possible to an explanation of the principles underlying the various parts of the brewing process, and have endeavoured in describing these to employ language which should be intelligible to the well-educated layman, for whom

PREFACE

this book is primarily intended. In dealing in so small a compass with a subject covering such vast areas in the domains of both biology and chemistry, the difficulty of knowing precisely what to include and what to omit has been very great, and this must be my excuse should any of my readers detect what they regard as errors of omission, or consider that I have submitted the subject to unduly great compression. The illustrations of brewing plant have been very kindly prepared for me by Messrs William Bradford and Sons, to whom my thanks are due. I am also indebted to Professor Percival and to Messrs Duckworth and Co., for permission to use the drawing representing the structure of the hop cone; to Mr Harold Wager, F.R.S., for permission to use the drawing representing the structure of the yeast cell, and to my assistant, Mr B. F. Sawbridge, M.A., for preparing certain of the photomicrographs.

A. C. C.

LONDON,
July, 1912.

CONTENTS

CHAP.		PAGE
I.	INTRODUCTION. Early references to brewing; different classes of beer brewed at the present day.	1
II.	MATERIALS USED IN BREWING. Malt and malting; grain-adjuncts; sugars of various kinds	8
III.	MASHING. Nature of the chemical changes involved; brewing waters; infusion and decoction systems	23
IV.	BOILING. Chemical changes which occur; hops—their botanical character, and more important chemical constituents	44
V.	COOLING. Coolers and refrigerators; bacterial and 'wild yeast' infection	62
VI.	FERMENTATION. Yeast—its biological characters—different species—'culture yeasts' and 'wild yeasts'; nature of the chemical changes involved in the process of fermentation; the enzymes of yeast; racking of the beer and cask-conditioning	73
VII.	DISTRIBUTION, INCLUDING BOTTLING. Bottling—different systems; by-products and their utilization	116
	APPENDIX	125
	BIBLIOGRAPHY	127
	INDEX	129

LIST OF ILLUSTRATIONS

	Longitudinal Section of Barleycorn	*Frontispiece*
FIG.		PAGE
1.	Mash Tun	27
2.	Fully grown hop cone	47
3.	Vertical Refrigerator	65
4.	*Saccharomyces cerevisiae*	74
5.	*Sacch. cerevisiae* budding	75
6.	*Sacch. cerevisiae* showing ascospores	77
7.	Structure of yeast-cell. 1. Nucleolus. 2. Peripheral layer of chromatin. 3. Chromatin patch on one side of nucleolus. 4. Nuclear vacuole. 5. Central volutin granule in the vacuole. 6. Chromatin network. 7. Granules of fatty substance. 8 Volutin granules. 9. Glycogen vacuoles. 10. Delicate suspending threads for the central volutin granule. (After Wager.)	78
8.	*Sacch. ellipsoideus*	85
9.	*Sacch. Pastorianus*	86
10.	*Sacch. anomalus* (film)	87
11.	*Sacch. apiculatus*	88
12.	Fermenting Tun	109
13.	Torula	119

CHAPTER I

INTRODUCTION

THE origin of beer, using that word in a general sense to indicate a fermented infusion of grain, is lost in the mists of antiquity. Probably the Egyptians were among the first civilized people to engage in the art of brewing, and there appears to be good reason for believing that barley wine or beer was well known in Egypt at least three thousand years before the Christian era. Herodotus, who is not always a model of trustworthiness, mentions that the Egyptians used a wine made from barley because there were no vines in their country, but this is clearly not correct since wine was well known to the ancient Egyptians, and its use is recorded as early as 4000 years B.C. It seems certain, moreover, that at that early period there were many vineyards in the Nile valley and that several kinds of wine were produced. Still it is not altogether unreasonable to suppose that in parts of the world where the grape would not grow, beer occupied the same position as wine in countries where the vine flourished. Among the ancient Hebrews beer was well known and although the word *Schekar*, used by Moses and occurring several times in the Pentateuch, might refer merely to strong drink in general, yet there seems to be good reason for supposing that it did in reality refer to an intoxicating drink prepared from barley. It may be mentioned in this connection that there is an old Rabbinical tradition that the Jews were free from leprosy during the captivity in Babylon by reason of their drinking Sicera (Schekar) made bitter with hops. If there be any truth in this, it is of special interest as showing at what an early period hops were used for flavouring purposes. But it was among the peoples of Western and North-Western Europe that beer was most largely consumed; and among the Gauls, Germans, Scandinavians, Celts, and Saxons, it had attained in very early times to the position of a national beverage. Some uncertainty attaches to the etymology of the words *ale* and *beer*, but there is very little doubt that the former (Saxon *Ealu*, Danish *öl*) is of Scandinavian and the latter probably of Teutonic origin, both words being used indiscriminately during Anglo-Saxon times. After a time the word beer appears to have dropped out of use in this country and was not again employed until about the fifteenth century, when the use of hops became general, the word being then applied to the hopped in contradistinction to the unhopped beverage, or ale. At the present day the two words are very largely synonymous, *beer* being used comprehensively to include all classes of malt liquor, whilst the word *ale* is applied to all beers other than stout and porter. Prior to the Roman invasion it is probable that *mead* or fermented honey was the beverage most commonly used in this country on festive occasions, and there can be little doubt that it is the most ancient of the intoxicating drinks of Western Europe. Next in point of antiquity came cider, and then with advancing civilization, beer. One of the earliest references to the use of ale in these islands is to be found in the *Senchus Mor*, which dates from the fifth century, and which shows that ale was well known in Ireland at the time of the arrival of St Patrick. In Wales too, about the same time, ale competed with mead as the drink of the wealthy. Among the Danes and Anglo-Saxons beer was certainly the favourite beverage, and its virtues are celebrated in many of their most ancient poems. It will be remembered that their conception of the highest felicity attainable by their heroes after death was to drink in the halls of Odin long draughts of ale from the skulls of their enemies slain in battle. It is not within the scope of this book to deal at any length with the history of beer, and it will suffice to say that from the fifth century onwards the popularity of beer increased to such an extent, that from being the occasional beverage of the wealthy, it had become in

the Middle Ages the general drink of all classes. The
ale most commonly consumed by the poorer classes
was doubtless low in strength and poor in quality,
and in this connection the following lines from
Piers Plowman are of interest. In speaking of the
independence of the labouring classes consequent
upon the scarcity of labour after the great plague in
the fourteenth century, Piers says :—

> 'Ne non half-penny Ale In none wyse drynke.
> Bote of the Beste and the Brouneste that Brewsters sullen.'

It is clear that at this time ales of several kinds or
strengths were brewed, and in old documents of the
thirteenth and fourteenth centuries the words *prima*,
secunda, and *tertia*, as applied to beer occur on several
occasions. Later, in the reign of Henry VIII, the
brewers were restrained from making more than two
kinds of beer, the *strong* and the *double*, and the prices
at which these were to be sold were fixed by Statute.
This restriction seems to have been withdrawn, for
later we again find that several varieties of beer were
produced and sold.

The ease with which the operations involved in
the production of beer from malted grain could be
carried out resulted in the installation of brewing
plant in the houses of vast numbers of the wealthier
classes, and almost every housewife of importance
may be said to have been her own brewer. During
the eighteenth century this practice fell largely into

disuse, and the brewing of beer passed for all practical purposes into the hands of the public brewer.

At the beginning of the reign of King William the duty on strong beer or ale was one shilling and threepence per barrel and the price charged by the brewer to his customers (who usually fetched it themselves from the brewery) was sixteen shillings per barrel. After our wars with France the duty was increased by ninepence per barrel on strong beer, and in 1694 it amounted to four shillings and ninepence on strong, and one shilling and threepence on light beer.

In the reign of Queen Anne the excise duty on malt, which had been originally imposed during the reign of Charles I but subsequently repealed, was re-imposed, and a tax on hops was levied, which remained in force until 1862. Towards the end of the seventeenth and at the beginning of the eighteenth centuries, the beer chiefly consumed in London was a mixture of heavy sweet ale with a lighter bitter beer, and a still weaker or small beer. About the year 1722 a beer was introduced which had the flavour and general qualities of a mixture of the three, and as its chief patrons were the labouring classes and porters, it became known by the name of *porter*. This beverage made rapid headway and in the early part of the nineteenth century it constituted the great bulk of the beer drunk in London. Thus during the year ending July 5th, 1812, no less than 1,318,037 barrels of porter were brewed in London by the twelve principal firms then in existence, whilst during the same period only about 105,000 barrels of ale were produced. In 1880 there was a complete revision of the mode of taxation to which brewers were subjected, the most important change being the replacement of the duty on malt by a duty on the finished beer, which was fixed at six shillings and threepence for each barrel of 36 gallons at a standard original gravity of $1057°$ less an allowance of 6 per cent. for unavoidable waste in the manufacturing operations. During succeeding years this duty was subject to many changes, and at the time of writing it amounts to seven shillings and ninepence per barrel of an original gravity of $1055°$, with an allowance of 6 per cent. for waste. By original gravity is meant the specific gravity of the wort prior to fermentation and it is on this that the brewer is taxed, the calculation being always made to a standard specific gravity of $1055°$. At the present time many different kinds of beer are brewed in this country, of which the more important are mild ale, porter, pale ale, bitter ale, and stout. In addition to these, which are sold both in cask and in bottle, there are less important varieties, peculiar to certain localities, as well as lager beers produced by the decoction system of brewing which is so widely adopted on the Continent and in America.

The great bulk of the beer made in this country is consumed by the working classes whose staple beverages are the so-called mild ale and porter. The mild ale is a lightly hopped beer of medium gravity and of full sweet flavour which in London is usually drunk when quite new, but which in the country is occasionally kept on draught for one or two months. Porter is a dark-coloured beer usually of about the same original gravity as the mild ale, and which is devoid of any pronounced hop flavour. Stout is a black beer of higher gravity than porter which it resembles in being somewhat lightly hopped. Its flavour varies greatly according to the locality in which it is sold, being sometimes (as in London) sweet and luscious, and at others (as in Ireland) devoid of any pronounced sweetness or *dry*. Bitter beers, as the name implies, are characterised by a marked flavour of the hop, and are brewed of several strengths. Pale ale, which includes ale intended for export and for storage, is the name usually given to beers of the highest quality. In original gravity the beers of this class vary very greatly, but they are always pale in colour and are almost invariably brewed from the finest materials, and represent in a sense, the highest product of the brewer's art. In connection with the above general descriptions, it should be pointed out that no sharp lines of demarcation can be drawn between any of the varieties mentioned.

These overlap very considerably, and in different parts of the country the same name is often applied to beers of very different characters.

The history of a beverage which in its various forms has played such an important part in our national and social life for so many centuries, is in the highest degree fascinating, and to those who are interested, I commend any or all of the historical works, the titles of which are given in the list on pages 127, 128.

CHAPTER II

MATERIALS USED IN BREWING

THE two essential chemical processes involved in the manufacture of beer are first, the conversion of starch by means of diastase into certain soluble substances, one of which is a fermentable sugar, and secondly the decomposition of this sugar by means of the yeast organism into alcohol and carbon dioxide gas. It is clear from this that the main raw material must be one containing starch, and that it should if possible contain the necessary converting agent, diastase. Both these conditions are satisfied in the case of certain germinating seeds, and although many have at various times been tried, long experience has shown that germinated barley possesses in the fullest degree the various properties necessary for the brewing of beer. Among other cereals which are occasionally employed are maize, rice, oats, and wheat, but these are always used, when used at all, in conjunction with a large proportion of germinated barley, or to give it its technical name, *malt*. Further reference will be made to these cereals later, and for the present I will confine myself to the consideration of barley malt which is the brewer's chief raw material. Although it is not within the scope of this book to deal at any length with the process of converting barley into malt, some reference to the subject is necessary to ensure a proper understanding of the mashing process, which is one of the most important parts of the whole brewing procedure. The process of malting may be said briefly to consist in artificially inducing the germination of the grain, and when sufficient growth has taken place, in stopping it by the application of heat. Simple as this bare statement may appear, the chemical and physiological changes occurring during the conversion of barley into malt are highly complex, and notwithstanding the vast amount of study devoted to their elucidation, are even now very far from being fully understood. Thanks, however, to the labours of Brown and Morris and others, the general character of these changes is tolerably clear. If a few corns of barley are bitten between the teeth they will be found to be very hard, whilst malt corns, tested similarly, will be found to be mealy and friable and to break down readily. Again, if a handful of barley be ground up and treated with a little warm water, no apparent change will occur, and even at the end of some hours the addition of a little iodine solution will give the blue colour characteristic of starch. If on the other hand some ground malt be treated in the same way, it will soon be evident that some chemical change is taking place, for the mixture, if fairly thick to commence with, will rapidly become more liquid, and the solution obtained after straining off the solid matter will be found to have a pleasant sweet flavour, and on adding a little iodine solution, there will be no blue coloration, showing that the starch has disappeared. It is clear then that some very marked changes have attended the transformation of the barley into malt, and these are of the utmost importance to the brewer. The above simple experiment shows that the malt differs from the barley from which it was made chiefly in containing an active substance which is capable under appropriate conditions of converting the starch of the grain into soluble, sweet-tasting products, and one object of the malting process is to produce these substances. But there is another and very important difference between the two. As has been pointed out above, the barley is hard and vitreous whilst the resulting malt is soft and friable, and a second object of the malting process is to bring about this change in the grain. A third object is to give to the malt the necessary flavour, which is accomplished by heating it to the requisite temperature on a specially constructed kiln. We will now consider briefly the meaning of the above-mentioned changes and how they are brought about by the maltster. A grain of barley consists essentially of two parts, the main starchy portion, known as the *endosperm*, and a smaller part at one end of the corn, known as the *embryo* (see frontispiece).

During the natural germination of the corn in the earth, the embryo, which as the name implies is the rudimentary plant, commences to develop, and the starch of the endosperm is the provision made by nature for the nutrition of the growing plant until such time as the first green leaves are formed and the rootlets have sufficiently developed to permit of their obtaining food from the soil. It is this process which the maltster imitates and turns to his own ends. In the earth the vital changes are initiated or developed by the absorption of moisture, and the first part of the malting process consists in soaking the grain in water until it has absorbed sufficient to start the growth. It is then spread on floors and the germination allowed to proceed until the plumule (frontispiece *c*) has proceeded about three-quarters of the entire length of the corn. It is during this period that the important change mentioned above takes place. In the barleycorn the reserve materials are not directly available for the nutrition of the growing germ, since both the starch and the greater part of the nitrogenous substances (proteins) are insoluble and consequently not diffusible through the cellular structures to the germ where they are required. In the barley itself, certain active nitrogen-containing substances occur known as *enzymes*, and others are developed during germination. These enzymes are the natural agents necessary for the conversion of the insoluble reserve food-material into soluble diffusible and assimilable substances, and we shall see a little later how necessary they are to the brewer in his operations. Between the embryo and the starchy endosperm there is a cellular body known as the *scutellum* (frontispiece *b*), the function of which is to secrete the necessary enzymes and which becomes active in this sense, as soon as germination has commenced. In addition to this, the *aleurone* layer (frontispiece *a*) is also active in the secretion of these enzymes. The starch of the grain is not in a free condition, but exists in the form of granules packed in innumerable vegetable cells, and the walls of these cells have to be attacked before the contained starch is capable of being acted upon by the enzyme *diastase*. This is effected by means of a second enzyme known as *cytase* which attacks the cellulose walls, either dissolving them or at least softening and rendering them more permeable to the diastase, which is then able to convert the starch into the soluble matters needed for the nutrition of the growing germ. This solvent or disintegrating action (its precise extent is still a matter of some uncertainty) of the cytase on the cellular structures of the endosperm explains the conversion of the hard and vitreous barley into the tender and friable malt. At the same time diastase is developed in considerable quantity, and a portion of the starch is used up by the young plant. It is, of course, to the interest of the maltster to reduce this consumption of starch to a minimum, as it is later on required by the brewer; and therefore, as soon as the action on the cell-walls, or 'modification' as it is technically termed, is sufficiently complete, which is usually the case when the plumule has grown nearly the length of the corn, further action is stopped by drying the grain, first on the malting floor itself and then on a kiln, either by the direct combustion products of coal or coke, or, as on the Continent, by means of hot-air. In addition to the solvent action of cytase on the cell-walls of the endosperm and the development of diastase so necessary for the subsequent conversion of the starch in the brewery, there are other changes occurring, one of which is of considerable practical importance. This is the conversion of a proportion of the insoluble proteins of the grain into soluble, diffusible and chemically simpler substances, such as proteoses, peptones and amino-acids. These, which are intended by nature for the needs of the growing plant, are of great importance to the brewer, as we shall see later, particularly in connection with the nutrition of the yeast during the process of fermentation. This protein conversion or *proteolysis* is effected by another enzyme or possibly by more than one, the action partly resembling that of *trypsin* and partly that of *pepsin* or *peptase*, the two enzymes present in the pancreas and stomach of animals, and which are responsible for the breaking down of the complex insoluble proteins during the process of animal digestion. Whilst only about 20 per cent. of the total nitrogen of barley is soluble, more than 40 per cent. of the nitrogen of the resulting malt can dissolve in water, the increase being chiefly due to the production of amides and amino-bodies such as asparagine. The changes taking place in the various nitrogenous bodies are illustrated by the following analyses of Bungener and Fries:

	Barley per cent.	Malt per cent.
Total nitrogen	1·690	1·580
Nitrogen as albumen	0·161	0·230
Nitrogen as peptone	0·040	0·060
Nitrogen as amides	0·206	0·534
Total soluble nitrogen	0·355	0·642

MATERIALS USED IN BREWING

The germinated grain, having been partially dried on the growing floor, is transferred to the kiln where growth is effectually arrested, and where it acquires a pleasant biscuity flavour.

Frequent use has already been made of the word *enzyme*, and perhaps it may be well to define a little more clearly what is meant, particularly having regard to the all-important part which these substances play in the processes of malting and brewing. The enzymes constitute a class of bodies which are of universal occurrence in all living organisms, whether animal or vegetable, and which are of supreme importance to the life and well-being of those organisms. Briefly their function may be said to consist in rendering soluble and available for nutrition the various insoluble reserve materials on which the life of the organism depends. Thus, we have starch-converting enzymes or *diastases*, cellulose-converting enzymes or cytases, sugar-converting enzymes such as *invertase* and *maltase*, protein-converting enzymes or *proteolysts*, fat-converting enzymes or *lipases* and finally the enzyme which splits up sugar into alcohol and carbon-dioxide gas (fermentation) or *zymase*. The enzymes are all nitrogen-containing substances of albumenoid character, and it is one of their most noteworthy characteristics that but very small quantities are needed for the conversion of relatively enormous quantities of the various substances on which they exert their specific actions. Technologically, this point is one of the highest importance, although the precise manner in which they exert their activity is not understood. They are soluble in water, and resemble certain of the proteins such as egg-albumen in being coagulated by heat. With this coagulation their chemical activity disappears, and this as we shall see is also a matter of the highest technological moment. The enzymes are not all equally sensitive to the destructive effect of heat, but all are destroyed in the presence of water at temperatures short of that of boiling water, and it may be said roughly that a temperature of about 80°—90° C. usually marks the limit of their activity. On the other hand they will withstand a temperature of 120° C. or even more when completely dry. They are capable of acting at fairly low temperatures, but for every enzyme there is a temperature or rather a limited range of temperature within which it is most active. Thus, the diastase of malt which converts starch into soluble sugar and other products functions most actively at about 50°—55° C. and the enzyme *invertase* which converts cane-sugar into invert sugar is most active at about 55° C. The enzymes are also exceedingly sensitive to the presence of very small quantities of many chemical substances, and whilst the majority act most readily in faintly acid solutions a few (e.g. trypsin)

MATERIALS USED IN BREWING

exert their specific activities best in faintly alkaline media

We have seen that during the conversion of the barley into malt the more important changes which occur are (a) the development of the enzymes (diastases) necessary for the subsequent conversion of the insoluble starch into soluble and partially fermentable products, (b) the modification of the cellular structures of the endosperm by means of the enzyme cytase, (c) the conversion of insoluble proteins into simpler and soluble nitrogenous products, and (d) the production by kilning of a pleasant biscuity flavour. Instead of the hard vitreous barley, we have now a mealy friable material containing a large proportion of the original starch (some is lost during malting by conversion into carbon-dioxide and water, as the result of the vital respiratory processes) and at the same time the enzymes necessary for effecting the important transformation of that starch which occurs during the process of mashing, and which will be dealt with in a subsequent chapter.

It has already been pointed out that it is one of the characteristics of the enzymes as a class that they are able to bring about an amount of chemical change which is enormously great in relation to the actual weight of the active agent concerned; and during the germination process a very much larger amount of diastase is formed in the growing grain than is needed for the brewer's purpose. One object of the kilning process is, therefore, to destroy, or to restrict the activity of a great deal of this diastase, leaving just as much in the finished malt as may be necessary for the purpose for which it is to be employed. This depends to a great extent on the class of beer for which the malt is to be used, a malt containing more active diastase (and therefore kilned at comparatively low temperatures) being required for pale ales than for mild ales or black beers. Even after the severe restriction of diastase which occurs on the kiln, barley malt contains much more diastase than is necessary for the conversion of its own starch, and in consequence other starch-containing materials may be used in conjunction with malt should their employment for one reason or another be deemed desirable. Such materials are maize, rice, oats, and, on rare occasions, other cereals. The maize and rice are usually submitted to a preparatory purifying and cooking process before being supplied to the brewer, and in the form of flaked maize and flaked rice, represent very pure and concentrated brewing materials, containing as they do from 75 per cent. to more than 80 per cent. of starch. The products obtained by the action of diastase on the starch of these materials are the same as those resulting from the conversion of the starch of barley-malt itself, but they yield to the worts much smaller proportions of soluble nitrogenous substances,

MATERIALS USED IN BREWING

and it is to this fact that their utility as malt-adjuncts is largely due.

Since alcohol is one of the most significant and important constituents of beer, it is clear that certain sugar products which are capable of yielding that substance on fermentation, might be used in place of a certain proportion of the malt. It will be seen later that during the mashing process the starch of the malt is converted partly into the fermentable sugar maltose, and that this is split up during the process of fermentation into alcohol and carbon dioxide gas. Since other sugars such as dextrose and laevulose undergo the same decomposition when submitted to the action of yeast, there is clearly no reason why a proportion of those sugars should not be employed by the brewer, assuming, of course, that in the form in which they are used they are perfectly wholesome and that they are found to yield good results in respect of the flavour and character of the finished beer. Cane-sugar itself is not directly fermentable, but is first converted by the enzyme, *invertase*, contained in the yeast cell into invert sugar, which then undergoes decomposition into alcohol and carbon dioxide. Cane-sugar may therefore be directly employed as a brewing material, but inasmuch as its use is thought by many brewers to conduce to yeast weakness, it is more usual to employ the invert sugar made from it. On the manufacturing scale the invert sugar is prepared by heating a solution of cane-sugar with a small amount of a mineral acid until the desired change is complete. The acid is then neutralized, and the solution, after more or less decolorization, is evaporated in a vacuum pan to the consistency of a syrup. In this process the cane-sugar undergoes hydrolysis and is converted into a mixture of dextrose and laevulose in nearly equal proportions, which is known as invert sugar. The change may be represented by the following equation:

$$\underset{\text{Cane-sugar}}{C_{12}H_{22}O_{11}} + \underset{\text{Water}}{H_2O} = \underset{\text{Dextrose}}{C_6H_{12}O_6} + \underset{\text{Laevulose}}{C_6H_{12}O_6}$$
$$\underbrace{\phantom{C_6H_{12}O_6 + C_6H_{12}O_6}}_{\text{Invert sugar}}$$

As used by the brewer invert sugar is a product closely resembling golden-syrup in appearance and in flavour, but when allowed to stand for some time it sets to a soft solid mass owing to the crystallization of the dextrose, the laevulose which crystallizes only with great difficulty, remaining in the syrupy condition. The commercial syrups usually contain about 75 per cent. of invert sugar, the balance consisting of water, with small quantities of cane-sugar, and a little mineral matter. In its composition, therefore, it is very similar to honey. The only other sugar product which is at all largely used in brewing is glucose. This is prepared by heating the starch obtained from various sources (maize, sago, potato, rice) with dilute

MATERIALS USED IN BREWING

mineral acid until the required amount of conversion has taken place. In this process the starch is converted first into maltose and dextrin, and finally, if the action be allowed to continue for a sufficient time, entirely into dextrose. The chemistry of this change is not, in point of fact, quite so simple as might appear from the above statement, small quantities of other and unfermentable carbohydrates[1] being formed at the same time. The action is usually allowed to proceed until the whole of the dextrin has disappeared, and the resulting product therefore consists substantially of a mixture of the two sugars, dextrose and maltose. Of course, in the process of manufacture, the acid used is neutralized and the solution decolorized as in the case of invert sugar. Thus prepared, glucose or starch-sugar forms a white or yellowish solid mass

[1] The term 'carbohydrate' is a generic term applied to a group of compounds, some of which are very widely distributed in nature, and which includes such important substances as cane-sugar, milk-sugar, starch and cellulose. The more important members of the group contain either 6 or a multiple of 6 carbon atoms in the molecule, and in all cases the hydrogen and oxygen atoms are present in the relative proportions in which they unite to form water. The expression 'carbohydrate' was obviously designed to call attention to the fact that the composition of these substances may be represented by the general formula $C_x(H_2O)_y$, and whilst the word is certainly not free from objection, it has at least the sanction of long and general usage, and is, after all, a convenient one by which to designate a group of important compounds closely related to one another and having many properties in common.

possessed of far less sweetness than invert sugar and containing about 65 to 70 per cent. of fermentable sugars.

By restricting the action of the acid in the manufacture of glucose, a product may be obtained containing a considerable proportion of dextrin and smaller quantities of dextrose and maltose. This product, which occurs in commerce in the form of a thick colourless syrup, is occasionally used in brewing, but finds its chief employment in the manufacture of confectionery.

In the brewing of stouts and porters, certain proportions of highly coloured malts and of caramel are employed. The malts are manufactured either by kilning the germinated barley at a high temperature over a fire of burning wood (brown malt) or by roasting barley or ordinary malt in a cylinder such as coffee is roasted in (black malt). These materials communicate to the beers in which they are used not only a considerable amount of colour, but also a pleasant and characteristic empyreumatic flavour. In place of a proportion of these coloured malts, caramel, prepared by the action of heat on glucose or cane-sugar, is often used.

With the important exception of water, hops and yeast (which will be dealt with more conveniently in subsequent chapters) the above may be said to constitute the materials from which beer is brewed.

CHAPTER III

MASHING

In a few words the object of this most important operation may be said to be the conversion by diastase of the starch of the malt or other grain (when this is used) partly into fermentable sugar, and partly into other substances which are either unfermentable or which only undergo fermentation with considerable difficulty. Simple as this statement may appear to be, the chemical changes involved in the process are exceedingly complex, and notwithstanding the immense amount of labour which has been for many years devoted to their study, it cannot be said that they are even now fully understood. Sufficient has, however, been learned in regard to the chemistry of the transformation of starch at the instance of diastase to furnish us with a tolerably clear insight into the nature of the changes occurring during the mashing process, and to serve as a reliable guide to the brewer in the control of his operations.

Before dealing with this part of the subject, it may be well to refer to the important bearing which the nature of the water used has on the quality of the resulting beer. At an early date, the town of Burton-on-Trent became celebrated for its pale ales, whilst Dublin and London became almost equally renowned for the excellence of their stouts and porters. Whilst this was to some extent due to differences in the methods of brewing adopted, it was soon recognized that it was chiefly owing to the different character of the water supplies of the respective districts; for even when the same materials and the same methods of brewing were employed, it was found for example to be impossible with the London water to brew pale ales having the same character and general excellence as those brewed in Burton, and on the other hand, Burton stouts and porters could not compete with those of London and Dublin. Now the water from the Burton wells is found to contain large quantities of calcium sulphate (about 80 grains per gallon) with smaller proportions of magnesium sulphate, and it is to the presence of these salts (particularly the former) in such comparatively large quantities that the special suitability of the Burton water for the production of pale ales is due. On the other hand the Dublin water contains little else than the carbonates of calcium and magnesium, and after boiling is consequently very soft, whilst the water from the deep wells in and around London contains sulphates, carbonates and chlorides of sodium, and is free from the earthy sulphates above referred to. Putting the matter very briefly it may be said that hard waters containing much calcium sulphate, such as those of Burton, are especially suited for the brewing of pale ales and the better class of bitter beers, whilst soft waters are best for the production of stout and porter. Mild ales and certain other classes of beer require for their production water intermediate in respect of hardness, between the two above-mentioned classes. Since the composition of the water necessary for the production in their highest excellence of the various classes of beer is well known, it will be clear that much may be done to convert an unsuitable water into a suitable one by artificial treatment, that is to say, by the addition of those mineral ingredients in which it is naturally deficient, or by the decomposition of those ingredients which are undesirable. Whilst the desired result cannot always be brought about by artificial treatment, it is possible in the great majority of cases to render certain water supplies much more suitable than they would otherwise have been, by the addition of the necessary materials such as calcium sulphate (gypsum), magnesium sulphate, calcium chloride, etc. Thanks to this knowledge it is now possible to brew pale ales of good quality in many towns other than Burton, and to use natural supplies which would otherwise be unsuitable for the production of certain classes of beer now largely in demand. It should be said that all water intended for brewing purposes, whether hard or soft, must be of a high degree of organic purity.

The first actual process within the brewery is the grinding of the malt. This is effected by passing the malt through revolving steel rolls, two pairs being often used side by side, with the rolls set at different distances, the small corns being passed through the one pair and the larger corns through the other. In this way, something approaching to uniformity of grinding is obtained. The object of the brewer is not to grind as finely as possible, but to crush each corn thoroughly so as to permit of the ready attack of the starch by the diastase when water is added, and yet to bring about as little disintegration as possible of the husk of the grain which is needed to assist filtration in the mash-tun. In certain modern processes of grinding the mealy portion of the grain is separated after crushing by various mechanical devices from the husk, and the former is then more finely ground, with the object of increasing the amount of extractive matter to be obtained from it; but it is not within the scope of this book to deal with working details, and in any case, the great majority of brewers still adopt the simpler though less perfect form of grinding machinery above referred to. The malt having been ground or crushed, the brewer is ready to commence the process of mashing. In this process, the crushed malt has to be mixed with the requisite amount of water under such conditions of temperature, that the action of the diastase on the starch can be kept within certain necessary limits. The vessel in which this action takes place is known as the mash-tun. As will be seen by reference to the following diagrammatic drawing (fig. 1) it consists essentially of a covered cylindrical vessel, constructed usually of wood or iron, fitted with a perforated false bottom (*a*), revolving rakes or stirring machinery (*b*), a sparging or washing appliance (*c*), and a number of pipes for

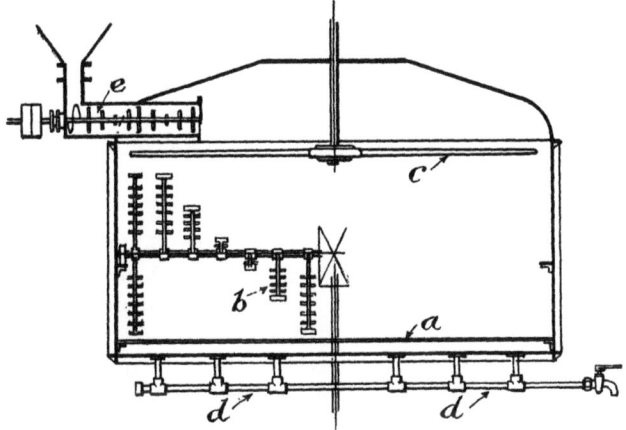

Fig. 1. Mash-tun.

drawing off the resulting clear wort (*d*). In some cases the admixture of crushed malt (grist) and water is made directly in the mash-tun itself, but more frequently a mechanical appliance known as an external masher (*e*) is used, by means of which a more perfect admixture of the malt and the water can be made, which enables the brewer to maintain a more exact control over the temperature of the mixture, which as we shall shortly see is a matter of the highest importance.

It will be remembered that during the process of malting, the whole internal structure of the barleycorn is altered in such a way that the starch granules with which the endosperm is packed are rendered more easily amenable to the action of the enzyme diastase when the crushed malt is mixed with water at suitable temperatures. It is generally held that this is due to the disintegration by enzymic activity of the walls of the vegetable cells in which those granules are formed and contained, and the process is technically referred to as *modification*. Ungerminated barley contains a form of diastase, the natural function of which appears to be the transference of starch from one part of the growing organism to another; this was termed by Brown and Morris 'translocation diastase' in contradistinction to the 'diastase of secretion' which is formed during the germination period. The precise chemical limitations and functions of these two forms of diastase are still uncertain; but it may be said that it is the 'diastase of secretion' (which will be subsequently referred to simply as diastase) formed during the malting process which is responsible for the conversion of the starch during mashing, and

MASHING

that this diastase is only capable of acting under the ordinary conditions of grinding and temperature on such of the starch as is contained in cells which have undergone the above-mentioned process of *modification*. From this it follows that if any portion of the barleycorn has escaped that change, that is to say, is hard and vitreous like barley, instead of being soft and friable like malt, it will in most cases escape conversion in the mash-tun and will so be lost to the brewer. It is usually stated as a general proposition that gelatinization of starch by heat must precede its conversion into soluble products by diastase, and whilst this is true of potato starch it does not appear to be the case with the starch of barley and certain other cereals. At any rate, in a well-made barley malt we have all the potentialities of the change which it is the object of the brewer to bring about. There is starch in a condition to be readily acted on by diastase, and there is more than sufficient diastase to convert that starch into soluble and partly fermentable substances. It will now be necessary to consider briefly the nature of that very important change. It may be said at once that the chemical reactions involved in the diastasic transformation of starch are of a highly complex character and are still the subject of investigation. It would obviously serve no useful purpose, even if it were possible within the limits assigned to this manual, to attempt to deal with the various views which have been held by the many chemists who have devoted themselves to this subject, and I therefore propose to state in the simplest language possible, the view which is most widely held at the present time, and which certainly offers the most satisfactory explanation of the observed facts. In regard to the composition of granular starch (that is, starch as it exists in various plants) we know nothing save that its 'empirical' or simplest formula is $C_6H_{10}O_5$. By submitting this to the action of cold dilute mineral acid, or in other ways, it may be converted into a simpler product which gives many of the ordinary reactions of granular starch, from which, however, it differs in being soluble in hot water and in not forming a paste or jelly when its hot aqueous solutions are cooled down. This substance is known as *soluble starch* and is the starting-point of the change which we are considering. It has the same empirical formula as ordinary granular starch, and there is some ground for assigning to it the molecular formula $(C_{12}H_{20}O_{10})_{100}$

When soluble starch is acted on for a long time by diastase at low temperatures, it is entirely converted into maltose, water entering into the reaction, according to the following equation:

$$C_{12}H_{20}O_{10} + H_2O = C_{12}H_{22}O_{11}$$

When, however, the action takes place at higher temperatures such as those adopted in the mash-tun, the starch molecule breaks down in such a manner as to yield maltose, dextrin and certain intermediate bodies known as *malto-dextrins* or *amyloins*. The soluble starch molecule may be represented as

$$\begin{cases} (C_{12}H_{20}O_{10})_{20} \\ (C_{12}H_{20}O_{10})_{20} \\ (C_{12}H_{20}O_{10})_{20} \\ (C_{12}H_{20}O_{10})_{20} \\ (C_{12}H_{20}O_{10})_{20} \end{cases}$$

At the moment of the attack by diastase this molecule breaks down into its five component groups, one of which differs from the remaining four in its resistance to the further action of the diastase, and constitutes the substance referred to in the literature of starch conversion as the 'stable dextrin.' The remaining four complexes, each having the formula $(C_{12}H_{20}O_{10})_{20}$, then undergo progressive hydrolysis[1], each $C_{12}H_{20}O_{10}$ or *amylin* group becoming converted by the assumption of water into a $C_{12}H_{22}O_{11}$ or maltose group. As each amylin group takes up the elements of water the resulting maltose group remains a constituent of the complex until the last has been hydrolyzed, when free maltose results.

[1] By hydrolysis is meant the conversion of one substance into one or more other substances of simpler molecular formula usually at the instance of dilute acids or enzymes, such conversion being preceded by the assumption of one or more molecules of water.

Thus, the change in the case of each of the four complexes might be illustrated by the following scheme:

$$(C_{12}H_{20}O_{10})_{20} \longrightarrow \begin{cases}(C_{12}H_{20}O_{10})_{19} \\ C_{12}H_{22}O_{11}\end{cases} \longrightarrow \begin{cases}(C_{12}H_{20}O_{10})_{18} \\ (C_{12}H_{22}O_{11})_2\end{cases} \cdots$$
$$\text{Malto-dextrin.} \qquad \text{Malto-dextrin.}$$
$$\cdots \longrightarrow 20 C_{12}H_{22}O_{11}.$$
$$\text{Maltose.}$$

From this it follows that if the diastase were allowed to act at favourable temperatures for a sufficiently long time, the final products of the reaction would be maltose and the 'stable dextrin' which is only acted upon with extreme difficulty. If, however, the temperature and time conditions are so arranged as to restrict the action (and this is the case in the brewer's mash-tun) then in addition to free maltose and the 'stable dextrin' a certain proportion of the intermediate products of the reaction, the malto-dextrins, will occur. It now remains to consider the parts which these various substances play in the production of beer. Beer differs from many alcoholic beverages in that its flavour and character are quite as much, if not more, dependent on the nature of the unfermented extractive matters, than on the presence of alcohol and other volatile products of fermentation. In addition to this, it is essential that it should be capable of undergoing a certain amount of fermentation while in the cask (and often in the bottle) awaiting consumption. This cask fermentation keeps the beer charged with carbon-dioxide gas, and without it the liquid would speedily become flat and undrinkable. It is clear from these considerations that the extractive matters derived from the malt or other materials must not be completely fermentable, for if that were the case, the resulting liquid would be little more than a dilute solution of alcohol, and would not possess any of the characteristics of beer. It may be mentioned in passing that this is the aim of the distiller, who desires only to produce alcohol, and whose methods are in consequence directed chiefly to that end. It will be remembered that the brewer makes use of malt which has been heated on the kiln to such a temperature as to bring about the destruction of much of the diastase formed during the germination of the grain, and in the mash-tun he again employs temperatures sufficiently high to restrict the activity of that which remains. In this way he ensures that the wort (that is the clear saccharine liquid containing the products of the conversion of the starch) shall contain (1) maltose, (2) malto-dextrins, and (3) stable dextrin. Now these substances together fulfil the conditions necessary for the production of beer, the maltose being readily fermentable and therefore yielding alcohol and carbon-dioxide gas, the malto-dextrins being less readily fermentable, and the stable dextrin practically unfermentable. The readiness with which the malto-dextrins are capable of undergoing fermentation depends to a great extent on their composition, those which contain the largest number of maltose groups in the complex molecule being hydrolyzed to free maltose and fermented much more easily than those containing a larger proportion of amylin groups. All, however, are capable of being converted into maltose by diastase. The brewer's wort in the fermenting tun does not, however, contain any active diastase, and for the present it will suffice to say that the great bulk of the malto-dextrins present in the wort are not fermented during the main fermentation by the ordinary yeast. Since maltose is completely fermentable and the stable dextrin for all practical purposes unfermentable, it follows that the quantity and nature of the unfermented, but fermentable, matter remaining in the beer at the end of the fermentation, will depend almost entirely on the proportion of malto-dextrins in the wort as it leaves the mash-tun, and on their character. This in turn is dependent on the nature of the malt employed and on the conditions to which it is subjected in the mash-tun. Broadly it may be said that malt kilned at a high temperature and mashed at a comparatively low one, will yield worts containing relatively high proportions of free maltose and of

MASHING

low-type[1] malto-dextrins, whilst malt kilned at lower temperatures and mashed at rather higher ones will give worts containing less maltose and a larger proportion of high-type malto-dextrins. The malto-dextrins, although behaving in some respects as actual compounds of maltose and dextrin groups, yet manifest many of the properties of mixtures. This is true of their flavour, that of the high-type compounds approximating to that of dextrin itself, whilst the 'low-type' malto-dextrins more nearly resemble maltose. In their degree of fermentability the same is true, for the 'low-type' compounds are obviously more easily converted into maltose, and therefore more readily fermented than those of higher type. In practice this is of considerable importance, particularly in connection with the brewing of beers of various classes. If we take a stock pale ale as representing the one extreme, and a mild ale intended for rapid consumption as representing the other, we

---

[1] By 'low-type' malto-dextrins is meant those containing a large number of maltose groups associated with a few amylin groups, and by 'high-type' those in which the number of amylin groups is largely predominant.

Thus $\begin{cases}(C_{12}H_{20}O_{10})_3 \\ (C_{12}H_{22}O_{11})_{17}\end{cases}$ and $\begin{cases}(C_{12}H_{20}O_{10})_2 \\ (C_{12}H_{22}O_{11})_{18}\end{cases}$

would represent 'low-type' compounds, whilst

$$\begin{cases}(C_{12}H_{20}O_{10})_{18} \\ (C_{12}H_{22}O_{11})_2\end{cases}$$

would represent one of 'high-type.'

shall see that the requirements are very different indeed. The former beer must be of delicate palate, free from excessive sweetness, and must contain a sufficient amount of residual carbohydrate matter to provide for a regular and constant supply of carbon-dioxide gas (cask fermentation or conditioning) over a considerable period of time. The mild ale, on the other hand, must be sweet rather than dry and must be in condition for drinking within often a few days of its manufacture. For the brewing of the pale ale, then, we require a wort containing a relatively large proportion of malto-dextrins of high type, since these are devoid of sweetness, and owing to the slowness with which under ordinary storage conditions they are converted into fermentable maltose, they supply the long-maintained and persistent cask fermentation which is a necessary feature of these beers. For the brewing of the mild ale, on the other hand, the wort must be rich in free maltose and in malto-dextrins of low type. Such wort will be readily fermentable and the beer at racking will contain carbohydrate matters sweet to the palate and capable of undergoing the quick cask conditioning which is essential in the case of beer intended to be consumed within a few days of racking. In practice it is found that these two sets of conditions are obtained by employing on the one hand a pale malt (i.e. a malt kilned at comparatively low temperatures) and adopting somewhat high mashing temperatures, and on the other, by using a higher-dried malt and mashing it at lower temperatures. It must not be thought that the range of mashing temperatures within which the brewer has to work is a wide one. So powerfully is the diastase affected by slight variations of temperature within the limits of practical working, that two or three degrees are often sufficient to affect to a marked extent the palate fulness, rate of conditioning, and other properties of the resulting beer. When working on the infusion system of mashing as adopted in this country, and excepting a few little-used processes, it may be taken that 145° F. and 155° F. represent the two extremes, the lower temperatures being employed in the brewing of mild ales, and the higher for pale ales, stock bitters and similar beers. It will be seen then that the great importance of the mashing process lies in the fact that it is in the mash-tun that the character of the resulting beer is largely determined, the brewer so arranging his conditions in relation to the malt, as to produce the type of wort, and consequently the class of beer he desires. Various methods have been devised for arriving at a knowledge of the amount of active diastase existing in malt, or rather for assigning to malts numbers, expressing in terms of an arbitrarily chosen scale, their relative diastasic activities. Whilst these methods do not bear any simple relation to the conditions of brewery practice, the results they yield are often of the greatest value, as showing whether any given sample of malt has been kilned in such a manner (that is to say whether its diastase has been suitably restricted) as to render it well adapted for the brewing of the particular class of beer for which it is to be used. The changes which have been described above and which occur when malt is mashed, apply equally to any 'raw' or unmalted grain which may be mixed in the mash-tun with the malt. As has already been pointed out, the latter contains more diastase than is necessary for the conversion of its own starch, and so a proportion of starch derived from some other source may be used if desired, the conversion taking place as in the case of the barley-malt starch itself. When, as is usually the case, some form of prepared grain such as flaked or gelatinized maize or rice is used, all that is necessary is to mix the grain uniformly with the malt in the mash-tun itself, and proceed exactly as if an all-malt mash were being made. In the manufacture of these flaked materials the grain is first submitted to a cleaning process and is then broken up by machinery into small pieces, in which form it is known as 'grits.' In the case of maize this is preceded by the removal of the germ, which contains the greater part of the oil, this being objectionable from the brewing point of view. The grits are then subjected to the action of steam whereby they are thoroughly softened, after which they are converted into flakes by being passed between hot-rollers, and finally dried. The main result of this steaming process is to break up the cell-structures of the grain and to gelatinize the starch and so render it amenable to the action of diastase at the comparatively low temperatures of the mash-tun. Occasionally, however, grain which has not been subjected to this 'cooking' process is employed, as for example, maize grits, broken rice, oatmeal (in oatmeal stout) and rarely barley. In that case the starch is not in a condition to be acted on by diastase under the ordinary mashing conditions, and it becomes necessary to submit the raw grain in question to independent treatment in a separate vessel known as a converter. In this vessel the grain, mixed with a small proportion of pale or active malt, is slowly heated with water to about 180° F. by means of steam, and is kept at that temperature for a little while, after which it is raised nearly to boiling. The object of mixing a small proportion of pale malt with the grain is in order that the liquefying diastase of the malt may convert the ordinary starch into the soluble form, and so prevent the grain mixture from becoming unduly viscous when cooled down to the temperature at which it has to be run into the mash-tun. The starch is now in a condition to undergo saccharification by the malt diastase, and the thin gruel is consequently run into the mash-tun with the malt grist and mashed in the ordinary manner. In this country the system of mashing known as the *infusion* system is almost universally adopted, but on the continent of Europe and in America a process of decoction is very largely employed. In the infusion system the ground malt, with other prepared grain, if such is used, is merely mixed with the requisite quantity of water at the most suitable temperature, and the action is allowed to proceed until conversion is complete and until wort of the required composition has been obtained, after which, the wort is run off, and the residual grains are sparged, the temperature rarely, if ever, rising above 165° F. In the decoction method, the malt is usually mixed with the water at a lower temperature than is customary in infusion mashing, and the temperature of the mash is raised by successive stages nearly to boiling. In some cases this is done by transferring a portion of the mash to a copper, and after heating it to boiling, returning it to the mash-tun, whilst in others a portion of the clear wort is run off so as to preserve a sufficient amount of active diastase, after which the contents of the mash-tun are raised to a high temperature by means of steam, and when cooled sufficiently the active wort is run in, and the mashing allowed to proceed. It will be seen that an important difference between these two systems lies in the fact that in the decoction method the mash is raised nearly to the boiling-point of water, and so any imperfectly modified grain is gelatinized and rendered amenable to the saccharifying action of the diastase. It follows that with ordinary malts, which, no matter how carefully made are apt to contain a little unmodified grain, the decoction system yields higher extracts, that is to say gives a rather greater proportion of soluble matter than the infusion. It is, however, very generally held that decoction mashing is not so well suited to the production of English beers as the method ordinarily followed, and in this there is some truth. The mashing, having been completed, the bright wort is run off from the insoluble portion of the grain, into the copper where it is boiled with hops. As the strong wort runs off, hot water is sparged on to the residual grains, so as to extract as completely as possible the whole of the soluble matter, and this too runs into the copper. The brewer, knowing the proportions of extractive matters yielded by the malts and other materials he is using, is easily able to calculate the total quantity of water necessary to produce the required volume of wort at the requisite density. The density of the wort, as will be seen later, determines the strength of the beer to be brewed, and as taken in the fermenting tun prior to fermentation, constitutes what is known as the 'original gravity' of the resulting beer.

The sparging process, to which reference has been made, consists essentially in washing out of the grains the soluble matters which have been formed during mashing. Care has to be exercised not to employ water at too high a temperature, since small quantities of unconverted starch left in the grains would be gelatinized and brought into solution, and as the small amount of diastase remaining might easily be rendered inactive, such starch would pass into the wort and might cause some difficulty in connection with the brightness of the beer. As a rule sparging commences at a temperature of about 170° F. which is somewhat in excess of the average mash temperature required, since some heat will have been lost during the period of standing, and this has to be made up. At the end of sparging the temperature of the residual grains, or 'goods,' to use the term usually employed, will be about 160° F. and the density of the last runnings, if the operations have been properly conducted, will not exceed 1002° or 1003°. In the case of strong beers, however when less sparging water has to be used this final gravity may be somewhat exceeded. The spent grains left in the mash-tun when sparging is finished contain a little starch, the amount depending on the thoroughness with which the malt had undergone modification during the malting process, and on the success with which the mashing operations have been carried out. They also contain the insoluble proteins of the malt, a little oil from the malt, and a good deal of digestible fibre. They have consequently a moderately high cattle-feeding value and are readily purchased for that purpose by farmers and others. Sometimes they are bought in their wet condition, with about 75 per cent. of water, but more often they are first dried, since in this condition they will keep good indefinitely, the wet grains rapidly becoming sour and unpleasant, particularly in the summer. The following may be taken as representing the average composition of dried spent grains as obtained in the ordinary infusion system of mashing:

Starch	4·15
Digestible Fibre and Gum	36·00
Fat and Oil	6·50
Albumenoids (Proteins)	18·40
Ash	4·12
Moisture	8·96
Crude Fibre	21·87
	100·00

Such a product has a combined feeding and manurial value represented by about 102 units, compared say with 104 for an average sample of wheat.

CHAPTER IV

BOILING

The wort from the mash-tun having been collected in the copper is ready to be submitted to the boiling process. Sometimes the wort passes through an intermediate vessel known as a receiver or underback from which it either flows by gravity, or is pumped into the copper. The use of such a vessel is determined very largely by the construction of the brewery and the relative positions of the mash-tun and copper. Technologically, the use of such a vessel has no special significance and the only point of any importance is, that the wort shall not be allowed to remain for any length of time in it at or below the temperature at which it leaves the mash-tun, since the diastatic action would, in that case, obviously be proceeding the whole of the time and a larger proportion of fermentable carbohydrate matter would be produced than might be required. In order to avoid this it is customary, in cases where the wort does not flow directly into the copper, to provide the receiving vessel with a steam coil by which the wort can be immediately heated to a temperature at which the diastase becomes inactive. Thus the wort when it reaches the copper will have practically the same carbohydrate composition as when it left the mash-tun. The copper may either be an open or a closed vessel capable of holding the whole or a part of the brewing. In the former case the whole of the wort is boiled in one operation but in the latter two or even three boilings may be necessary, the stronger wort being boiled first and the weaker wort subsequently. The copper is boiled either by fire or by steam, and as a rule a boiling period of about two hours is adopted. The objects of the boiling process are the following :

    (a) The sterilization of the wort.
    (b) The arresting of the action of the diastase.
    (c) The extraction of the flavouring and preservative constituents of the hops.
    (d) The precipitation of undesirable protein matters.
and (e) The concentration of the wort to the requisite point.

It will probably be conducive to clearness if the above five objects are dealt with separately, but before doing this it may be convenient to devote a little space to a description of the hop plant or at least to that portion of it which is used by the brewer.

The hop belongs to the *Cannabaceae* but it possesses certain affinities with the stinging nettle and is in consequence occasionally classed with the

Urticaceae. It is probable that hops were grown, and used chiefly, perhaps, for medicinal purposes at a very early period. Occasional references to hops and hop gardens occur in documents of the ninth century and it seems not improbable that even at that early period they were occasionally used for the bittering of beer. By the thirteenth century the area under cultivation had apparently increased very considerably and in the fourteenth century there is plenty of evidence that hops were employed for the bittering of beer, at any rate in Germany and in Holland. It is generally supposed that hops were first introduced into England towards the close of the fifteenth century, but that they were not received with open arms is evidenced by the fact that both Henry VII and Henry VIII prohibited their use in beer. This ban appears to have remained in force until the reign of Edward VI when the restrictions as to the employment of hops in brewing were removed and their cultivation was very considerably extended. The common hop is dioecious,—that is to say, the male and female flowers are produced on separate plants. The female flower which alone is used by the brewer consists of a cup-shaped corolla with a round ovary containing one seed. A considerable number of these flowers grow together in the form of cones which are technically known as strobiles. It is these strobiles which constitute the

IV] BOILING 47

hop as used by the brewer and which contain the various constituents which are of so much importance. The following illustration, fig. 2, represents the structure of one of these strobiles.

No. 1 shows a fully grown strobile consisting of an axis or strig on which are arranged bracts of two

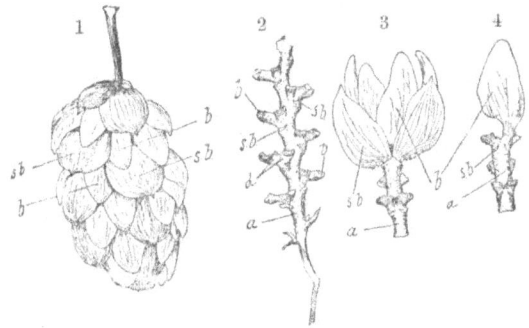

Fig. 2. 1—Fully grown hop cone; *sb.* seedless bract or 'petal' *b.* seed-bearing 'petal.' 2—Hop axis or 'strig.' 3 and 4—Pieces of strig and petals as in 1.

different kinds. The one class of bracts (*b*) contains the fruit or seed of the hop, whilst the other (*sb*) is seedless. The hop axis or strig is shown in No. 2, whilst No. 3 and No. 4 show the manner in which the bracts are attached to the strig. At the bases of the bracts will be found a yellowish powder known as lupulin, which, when examined microscopically, is found to consist of granules of regular shape and well-defined structure. This yellow powder, which is usually spoken of by the brewer as 'condition' is of the greatest importance since it contains the bulk of the constituents on which the brewing value of the hop chiefly depends. Other things being equal, therefore, the commercial value of a sample of hops is roughly proportional to the amount of lupulin which it contains. The percentage of lupulin in different hops is subject to very considerable variations. In old hops it may be as low as 2 or 3 per cent., whilst in new and rich hops it may occur to the extent of 16 per cent. or more. It contains the essential oil, resins, wax, bitters and a number of nitrogenous bodies including one or perhaps more alkaloids. Of these constituents the essential oil, certain of the resins and the bitter principles are of special importance, the essential oil and the bitter substances being the chief flavouring constituents, whilst certain of the resins are markedly bactericidal and so confer on the hops their well-known preservative properties. The essential oil occurs to the extent of about 0·3 to 0·6 parts per 100 parts of the hops, or rather it should be said that this is the amount that can be obtained from the hops by distilling them with steam, and recovering the oil from the aqueous distillate. The oil consists, as the author has shown, of two hydrocarbons, *myrcene* and *humulene* (inactive caryophyllene) and

BOILING

of several oxygenated substances which are present in comparatively small proportions, but on which the odour of the oil is largely dependent. These oxygenated constituents vary somewhat in oils of different origin, and it is due to this variation that a sample of oil obtained from say Californian hops differs appreciably in odour from one prepared from Bavarian or Kent growths. The hydrocarbon myrcene is a very mobile liquid having a penetrating and not unpleasant odour, and undergoes conversion into a non-volatile resinous substance on exposure to the air. As this change takes place with great readiness and as the myrcene constitutes about 40 to 50 per cent. of the fresh oil, it will easily be understood that the yield of volatile oil from hops a few months old is very considerably less than from the same hops when freshly picked. The humulene is a representative of the class of substances known as sesquiterpenes and when pure possesses very little odour, nor does it undergo any appreciable change when exposed to the air. It is usually present in fresh oil to the extent of about 40 per cent. The oil is almost insoluble in water (about 1 part in 20,000 parts), but it dissolves a little more readily in a weak alcoholic liquid such as beer. Slight, however, as its solubility is, it is yet ample for flavouring purposes, for, as with most essential oils its odour and flavour are most apparent when it is in a highly diluted condition. I have pointed out that it is volatile with steam and as might be supposed it is very largely lost when the hops are boiled with the wort, almost the whole passing away into the air with the escaping steam. The fragrant smell in the neighbourhood of a brewery when the wort is being boiled affords some evidence of this. To minimise the effect of this loss most brewers are in the habit of adding a proportion of the best hops to the copper shortly before the conclusion of the boiling period. It appears very probable, however, that the flavour (other than bitter) communicated to the wort by the hops is due not so much to the volatile oil itself, as to the solution of a small quantity of the resinous oxidation products of the oil, which are not volatile with steam, and which possess a smell and flavour very similar to those of the oil itself. In certain classes of beer it is customary to add a small quantity ($\frac{1}{4}$ lb. or $\frac{1}{2}$ lb.) of hops to the beer in the cask, and in this case the oil will, of course, play a more important part, as some of it passes into solution, and so communicates to the beer its characteristic flavour and aroma. It may be added that unlike many essential oils, the oil of hops does not possess any antiseptic properties.

We now come to a brief consideration of those highly important constituents, the resins and bitter substances, which are not only flavouring agents, but which exercise the even more important function

BOILING

of preserving the beer from the deteriorating effects of bacterial activity. The chemistry of these substances is still lamentably incomplete, notwithstanding the large amount of work which has been devoted to their study. Three distinct resins have been up to the present isolated, known respectively as the α-, β- and γ-resins, the two first being possessed of antiseptic properties, and the last being devoid of any such power. For technical purposes it is customary to distinguish merely between the so-called 'soft' and 'hard' resins, the former being soluble, and the latter insoluble in light petroleum. It is to the soft resins that the preservative properties of the hop are ascribed. This division is not a very scientific one, for both classes of resin unquestionably consist of a number of substances about which little is known. It has nevertheless the advantage of being convenient, and does to some extent at least connote a difference which is of technological importance. The soft resin is an exceedingly unstable substance, and tends to pass very readily into the hard. This change occurs in the hops during ordinary storage, and it is to this, that the greatly reduced preservative value of old hops is due. It has been found that when hops instead of being kept at ordinary atmospheric temperatures are stored at temperatures between 30° F. and 40° F. the activity of the various chemical changes which produce such marked deterioration is greatly reduced. The following table, due to Briant, shows for example the influence of temperature on the proportions of soft and hard resins in the same hops, when stored for the same time:

	Soft	Hard	Total
Hops as put in bottle	11·75	3·16	14·91
A. Hops stored seven months at 72–75° F.	8·82	5·94	14·76
B. Hops stored seven months at 55–65° F.	9·21	5·15	14·36
C. Hops stored seven months at 35–40° F.	10·67	4·20	14·87
D. Hops stored seven months at below 32° F.	11·10	3·57	14·67

At the present time, very large quantities of hops are cold stored as soon as bought, and the brewer is then able to avail himself late in the season of a material possessing preservative properties but little inferior to those of the original hop. In close genetic relationship with these resins stand certain substances of acid character, several of which have been obtained in a crystalline condition. These are the so-called hop-bitter acids. These substances, like the resins, are characterized by great instability, passing readily into resins, either on exposure to air or when boiled with water. One of these acids yields mainly valeric acid on oxidation, and it is to this that the cheesy odour of old hops is due. Since the preservative properties of hops do undoubtedly reside in the so-called soft resins, many attempts have been made to regard the percentage of these resins as the basis of a chemical evaluation of hops for brewing purposes.

BOILING

Speaking in general terms, it is unquestionably true that those hops (e.g. high class Bavarian, Californian, etc.) which contain the largest proportions of soft resin as determined by extraction with light petroleum, are those which practical experience has shown to have the strongest preservative properties in practice, and there can be no doubt that the method is, within certain limits, a very useful one. It cannot be denied, however, that it has its limitations, and much work will have to be done before the precise parts played by the various resins and bitter acids in the preservation of beer is fully understood.

It has already been stated that the resins occur almost entirely in the lupulin of which they constitute from 50 per cent. to 70 per cent. or more. Hops grown in different countries differ not only in the percentage of resin which they contain, but also in their aroma, which depends on the precise composition of the essential oil, a fact which often influences the brewer in selecting the blend of hops to be used. Hops like very many other plants contain tannin, in amount varying from about 3 per cent. to 5 or 6 per cent. At one time considerable importance was attached to this constituent, since it was thought to be a potent factor in the coagulation and subsequent removal of undesirable protein matters from the wort. That it is operative to a small extent in this direction is true, but there is no ground for supposing that any definite connection exists between the richness in tannin of hops and their value to the brewer. In addition to the above constituents hops contain gum-like bodies and a number of nitrogenous compounds the precise significance of which in brewing technology is not yet completely understood. It had long been known that the addition of a small quantity of fresh hops to beer in cask was usually followed by an outburst of fermentation, a fact which did not receive an adequate explanation until Brown and Morris showed that like most plants they contain diastase, which, of course, converts some of the malto-dextrin present into readily fermentable maltose.

With this brief account of the more important constituents of the hop, it will now be possible to understand the general character of the changes occurring during the boiling process, and it will perhaps conduce to clearness if we consider seriatim the five objects given on page 45. The first, and not the least important is the sterilization of the wort. It will readily be understood that the wort when it reaches the copper contains enormous numbers of living organisms of many kinds (chiefly bacteria, yeasts, and moulds) derived from the surface of the malt and other materials used in the mash-tun, the mashing temperatures being as a rule too low to effect the destruction of vast numbers of such organisms. Although hopped-wort does not constitute a very favourable medium for the development of many of these, and beer is still less favourable, yet those capable of developing would be sufficiently numerous to render the finished beer bad and undrinkable in a very short time. During the boiling period, however, these are all destroyed and the wort when it leaves the copper is perfectly sterile, that is it contains no living organisms of any description. The second object of the boiling process is to arrest the action of the diastase. It has already been pointed out in the previous chapter that the character of the beer produced is very largely dependent on the nature of the carbohydrate constituents of the wort when it leaves the mash-tun, and that this in turn is dependent on the extent to which the diastase of the malt has been allowed to act upon the starch. This part of the process is capable of being closely controlled by the brewer, who is able so to adapt the conditions to the material to be used, as to produce a wort having just the degree of fermentability required. Since the diastase is not destroyed at the temperatures of the mash-tun, but retains much of its activity in the wort, it is clear that unless steps were taken to arrest this activity, the conversion process taking place during mashing would continue, with the result that unduly large proportions of maltose would be formed and the resulting product, after fermentation, would be merely an alcoholic liquid, with little or none of the reserve carbohydrate material needful for cask fermentation, and possessing none of the characters of beer. On boiling, however, this diastatic activity is completely arrested, and so the composition of the wort is fixed. In regard to the third object mentioned on page 45, namely, the extraction of the flavouring and preservative constituents of the hops, it will not be necessary to say very much, in view of the description of those constituents which has been given above. From what has been said, it will have been gathered that the bittering and preservative constituents of the hop are not clearly distinguishable, but that both sets of properties are resident in the bitter acids and resins. Owing to their great instability these substances are quickly decomposed in the copper, becoming largely converted into antiseptically inert substances, and consequently their subjection to a prolonged period of boiling is objectionable. On the other hand the full bittering effect is not so easily obtained, and consequently the brewer is compelled to adopt a procedure which is in the nature of a compromise, that is to say some of the hops are added at the commencement of boiling, the remainder (usually the best) being added shortly before the completion of the process. In this way it is probable that the best results are obtained from the hops, both in respect of preservation and flavouring. Reference has been already made to the advantage of this procedure in reducing to some extent the loss of essential oil. The quality of the hops and the proportions used will naturally depend upon the class of beer to be brewed. Thus in the brewing of stock pale ales and bitter beers larger quantities of hops, and those of superior kinds, will have to be used, than in the case of mild beers. The reasons for this are twofold. In the first place the former beers must have more hop flavour than the latter, and in the second, they are usually expected to remain sound for a much longer period and consequently need more of the preservative constituents. In beers brewed for export, this is particularly the case, since such beers are often exposed to a very great strain, as for example in tropical countries, and in their production large proportions of the best and strongest hops are invariably employed. It may not be out of place to refer here to an important difference between the great bulk of the beer brewed in this country, and the lager beer of the Continent and America. The latter beer, when brewed, is kept for a considerable time (often some months) in casks stored in cellars kept nearly at the freezing point. When this lagering process is complete the beer is transferred to the trade casks, and must be quickly consumed if it is to be drunk at its best. In the brewing of these beers the Continental or American brewer uses a much smaller proportion of hops than is usual in this country for stock beers, the reason being that whilst the English brewer has to rely almost entirely on the hops for the keeping of his beer, the lager beer is preserved by being kept at very low temperatures.

We may now consider the fourth object of boiling, that is the precipitation of undesirable protein matters. The mash-tun wort contains nitrogenous substances of many kinds. Together with such comparatively simple substances as amino-acids and amides, it contains soluble and non-coagulable products of protein hydrolysis, such as proteoses and peptones, and finally more complicated proteins which though soluble in the wort at the temperature at which it leaves the mash-tun, are converted into insoluble substances (coagulated) on boiling. It has already been pointed out that during the malting (germination) process the proteins of barley undergo considerable change, being converted by the proteolytic enzymes of the grain into simpler and more soluble products. The resulting malt, therefore, contains a much larger proportion of its nitrogen in the form of soluble and non-coagulable products than is the case with the barley from which it has been made, and this process of enzymic hydrolysis and simplification proceeds during the mashing process. Many of these simpler nitrogenous substances are necessary for the nutrition of the yeast during the subsequent process of fermentation, and their presence in the wort in sufficient quantity is therefore of the highest importance. The more complex proteins, however, are so far as is known useless for the purpose, and as their presence in the finished beer could only have the effect of seriously diminishing its keeping properties (to say nothing of its brilliancy), their removal is in the highest degree desirable. Fortunately these proteins are rendered insoluble on boiling, and can, therefore, be readily removed when the boiled wort is subsequently filtered in the hop-back. The last object of the boiling process is to effect the requisite concentration of the wort. In order to wash out of the grains in the mush-tun the whole of the soluble saccharine matters it is necessary to employ somewhat considerable volumes of water, and some of this has to be removed by evaporation in order to reduce the wort to the necessary volume and gravity. A moment's consideration will show that in the brewing of say one hundred barrels of beer, it is necessary to use much more than one hundred barrels of water, quite apart from that required for washing purposes and for cooling. In the first place the spent grains left in the mash-tun retain a considerable quantity amounting to nearly 30 gallons for every quarter of malt mashed. Then there is the loss by evaporation in the copper and subsequently during cooling, the quantity retained by the hops in the hop-back, and finally there is a small loss due to the transference of the wort from one vessel to another. Speaking roughly, it may be said that in the actual brewing of one hundred barrels of beer about one hundred and thirty barrels of water would be required. Brewers' coppers vary a good deal in construction, the majority being open, whilst others are closed and so constructed that the wort can be boiled under slight pressure. The effect of the increased pressure is, of course, to raise the boiling-point of the wort a few degrees, and to bring about the extraction of rather more matter from the hops, than would be the case with an open copper. It may be doubted whether this is ever desirable, and the general consensus of opinion is undoubtedly in favour of boiling under ordinary atmospheric pressure. Boiling by steam is largely replacing the older method of fire-boiling, being cleaner, more convenient and more economical. When the wort has been boiled for the requisite time (usually about two hours) it is allowed to flow from the copper into a wooden or metal vessel known as a *hop-back*. This may be of any convenient shape (usually round or rectangular) and is provided with a false bottom consisting of a series of perforated metal plates. The object of this vessel is to retain the spent hops, and to allow the wort to be drawn off in a fairly bright condition. Since the hops absorb a good deal of wort, it is usual either to wash this out by sparging with hot water, as after mashing, or to obtain it by means of a suitable press, to which the residual spent hops are transferred. The layer of hops which covers the false bottom acts as an excellent filtering medium, keeping back the coagulated proteins referred to above, and enabling the wort to pass to the next stage in a purer and brighter condition. From this point of view it is obviously desirable that the area of the hop-back should not be too large in relation to the volume of wort to be dealt with, so that a layer of hops sufficiently deep for good filtration may be obtained. It may perhaps be mentioned here that the quantity of hops used in the brewing of beer in this country varies from about 1 lb. per barrel of beer in the case of mild ales to four or five times that quantity in the case of fine pale ales, strong stouts, and certain export beers. When sugar materials (invert sugar, cane-sugar, glucose, etc.) are used, it is customary to dissolve them in a separate vessel and to run the solution into the copper, or they may be added directly to the wort in the copper itself. In Chapter III reference has been made to the fact that the various classes of beer require for their production water of different mineral character, if the best results are to be obtained. Taking on the one hand a soft alkaline water such as is obtained from the deep wells in and around London, and on the other a hard gypseous water such as that derived from the wells in Burton-on-Trent, it may be useful to refer here to certain differences which these waters exhibit and which are of importance from the brewing point of view. In the first place, the softer water will prove rather more extractive in the mash-tun, and also in the copper. From the hops it has a tendency to dissolve certain coarse and unpleasant bitter substances which are not dissolved to the same extent by the harder water, and it also produces wort and beer of rather higher colour. Worts, moreover, which have been brewed with the hard gypseous water 'break' better on boiling, than those brewed with a soft alkaline supply; that is to say the coagulable proteins which are rendered insoluble during boiling, form larger and more coherent masses, and so are removed more completely during the hop-back filtration mentioned above.

CHAPTER V

COOLING

THE next stage in the conversion of malt into beer is the cooling of the wort to the temperature at which the yeast is added and fermentation commences. Simple as this statement may appear the cooling of the wort is one of the most important and, in some respects, most troublesome of all the

COOLING

procedures which go to make up the brewing process. The mere lowering of temperature presents, of course, no difficulty, but it has to be remembered that wort, even when hopped, is not an unfavourable medium for the development of a variety of living organisms, which as the result of their life activities bring about chemical changes which may render the resulting beer unpleasant in flavour or even quite undrinkable. As will be seen later such organisms may be either bacteria or certain species of yeast, and the chief aim of the brewer at this stage is to cool his wort to the required temperature and at the same time to protect it from infection by such undesirable organisms which are always present in the air. As a general rule, the cooling process takes place in three stages. In the first place there is a small reduction of temperature during the short time the wort is in the hop-back, then there is a further considerable lowering of temperature on the cooler, and finally the temperature is reduced to the required point by being passed over one or more refrigerators.

The cooler is a shallow rectangular vessel made of iron, copper, or wood, and is usually situated near the top of the brewery and in such a position that air may have free and ready access to it. As this vessel is almost invariably at a much higher level than the hop-back, the wort is pumped from the latter vessel on to it and allowed to remain as a rule until its temperature has been sufficiently reduced to permit of its being run over the refrigerator Since on the cooler the wort is only a few inches deep and consequently presents a very large surface to the air, it is clear that under ordinary circumstances immense numbers of organisms of various kinds must fall into it. As a rule the room in which the cooler is situated has open louvre boards on all four sides, so as to allow as much air as possible to pass over the cooling wort and to facilitate the escape of steam. As, moreover, breweries are usually built on or very near to main roads, it will easily be understood that the air which obtains access to the wort, particularly during dry, dusty weather, is very rich in all kinds of micro-organisms. Prior to the introduction of refrigerators, the wort had to be cooled entirely on the cooler, which frequently necessitated a sojourn of 24 or 36 hours, particularly during the summer months. It will readily be understood that under such circumstances it was often impossible to brew beer possessed of reasonable soundness, and so notwithstanding their greater alcoholic strength as compared with the beers of to-day, and the larger proportions of hops used in their production, they were frequently sour and undrinkable, almost at racking. In fact it was found that beers which had to be stored for any time could only be brewed during the winter months, and even then their stability was often very problematical. The introduction of refrigerators marked an epoch in brewing practice, since it enabled the brewer to cool the wort to the required temperature in a very short time, and with a minimum exposure to the air. Re-

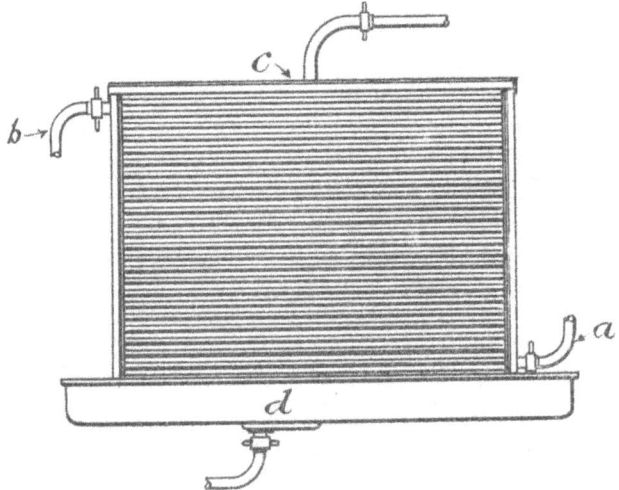

Fig. 3. Vertical Refrigerator.

frigerators are ordinarily of two forms, the vertical and the horizontal. In both, however, the principle is the same, the wort running over metal tubes through which cold water is made to flow. The cold water in the vertical form (fig. 3) is admitted to the lowest tube at (*a*) and thence passes successively to the others, and finally flows away at the top (*b*). The wort passes from the trough (*c*) through a series of small holes in a thin film over the outer surface, and when it reaches the pan (*d*) has been reduced to the temperature necessary for the addition of the yeast. In the majority of breweries at the present day, both a cooler and a refrigerator are employed, but sometimes the cooler is done away with entirely or is replaced by a deep receiving vessel, a system which has much to recommend it, as I shall show later. It should be said, however, that the cooling of the wort is not quite the only function of the cooler Certain of the constituents of the wort have the property of absorbing oxygen from the air at tolerably high temperatures, and this 'hot aeration' as it is called, to distinguish it from the cold aeration or absorption of oxygen by the cold wort while passing over the refrigerator, is very generally regarded as beneficial. It is true that some authorities have questioned its importance, but I think there is a general consensus of opinion that these more or less obscure oxidation changes are desirable and that they do exert an appreciable effect on the brightening capacity of the finished beer. It is fortunate, however, that these changes occur most actively at elevated temperatures,—about 180° F., and it is very doubtful whether much if any advantage in this direction is gained by

COOLING

allowing the temperature of the wort on the cooler to fall below,—say 160° F. A further function of the cooler is to permit of the deposition of the coagulated protein matters from the cooling wort and to leave the bulk of these behind when the wort is run down over the refrigerator. This insoluble deposit is technically known as the 'cooler sludge.' There can be no doubt that if the wort on the cooler could be reduced to a comparatively low temperature under conditions rendering bacterial infection impossible, the shallow vessel of large area has much in its favour. In practice, however, this is very difficult, and many brewers have found that they can secure the main benefits of a cooler without its serious drawbacks by substituting for it a deeper vessel of much smaller area. By spraying the wort pumped from the hop-back into such a vessel sufficient 'hot aeration' is secured, and if it is not possible to keep the protein sludge back as completely as with a cooler, the advantages on the score of diminished infection are so great as to render this consideration of little importance. Above 150° F. the wort is practically sterile, and the brewer should endeavour by every means in his power to ensure that the temperature of the wort when it reaches the refrigerator shall not be appreciably lower. Assuming the wort to be for all practical purposes sterile when it reaches the refrigerator, the next consideration is how to avoid serious infection while the wort is passing over that piece of plant. It has been pointed out that the wort runs over the refrigerator tubes in a thin film (this is clearly necessary in order to secure rapid cooling) and it will be obvious that the surface exposed to the air by a brewing of say a hundred barrels must be enormous. The adequate protection of the wort at this stage does not, however, present any very serious difficulty, since it is not a difficult matter to enclose the refrigerator, or refrigerators, in a tightly constructed room capable of being supplied with filtered air, and this is the procedure adopted in most modern breweries. The purification of the air from micro-organisms is effected in many ways, such as by passing it over numerous trays containing jelly or by causing it to traverse cotton wool or other filters, the filtering material being usually kept moist in order to increase its efficiency. The air so filtered is then injected by means of a fan into the refrigerator room, and in this way the steam is removed and the surface of the cooling wort is brought into contact with a constant supply of pure air. Whilst it is necessary to protect the cooling wort from *air-borne* infection, it is also of the highest importance that the refrigerator itself should be in a scrupulously clean condition, and the efforts of brewers' engineers have during recent years been directed to so improving the design and construction of this important appliance as to render cleaning a simple operation, and the lodgment of dirt practically impossible. Unless the greatest care is taken dirt is very liable to accumulate in the corners where the tubes enter the upright supports, and as this will always contain innumerable living organisms, it is quite easy for a whole brewing to be infected and spoiled by being passed over a refrigerator which is not properly clean. It will be seen then that infection of the wort may occur either on the cooler (whenever the temperature falls appreciably below 140° F.) or on the refrigerator, and may be caused either by air-borne organisms or by those derived from contact with dirty surfaces. As showing the amount of infection which may be caused by a refrigerator which has not been thoroughly cleaned, I may say that I have on many occasions found the wort at the top of the refrigerator practically sterile, whilst samples taken at the same time from the pan have contained as many as 3,000,000 organisms of various descriptions, per litre. Fortunately numbers such as these are not as terrible as they may at first sight appear to be, since hopped wort and beer are not very favourable media for the growth of bacteria in general. Thus, Zikes has shown that of 107 varieties of bacteria experimented with (including bacilli, micrococci, sarcinae, and spore-forming bacteria of various kinds) only 15 per cent. could develop in hopped wort, provided that the wort was simultaneously seeded with yeast, and that only 2 per cent. were capable of growing in beer at 10° C. and 3·7 per cent. in beer at 25° C. On the other hand, it must not be forgotten that bacteria reproduce with enormous rapidity under favourable conditions, and that the organisms which are capable of bringing about disease-changes in beer are of very widespread occurrence. Zikes' experiments, which dealt more particularly with water bacteria, are in accord with some of my own observations in connection with air-borne bacteria, for I have observed that in many cases the proportion of bacteria capable of developing in hopped wort does not exceed 5 per cent. of the total number falling in. This fact points to an important difference between infection due to chance air-carried organisms and that caused by dirty surfaces such as I have referred to above. In the latter case the organisms which have survived as the result of a kind of natural selection are all capable of growing in hopped wort and many of them in beer. They are in fact all effective, and a given number thus introduced is almost certain to have a much greater effect on the wort and beer than a similar number falling in from the air. It must be remembered, moreover, that bacteria are not the only organisms which the brewer has to fear, since some species of yeast are pathogenetic to beer, and these flourish without exception in both wort and beer.

These so-called 'wild yeasts,' to distinguish them from the industrially useful or culture yeasts, are usually present in considerable numbers in the air from June to September, after which date they rapidly diminish. Growing as they do on the surface of ripe fruits such as cherries, plums, strawberries, etc. it will be obvious that wort infection by these yeasts is more likely to be serious in the case of breweries situated in the country and in the neighbourhood of orchards and fruit gardens, than in those in towns, although this is not by any means always the case. Since the thermal death-point of these yeasts is usually about 140° F. it follows that no infection is to be feared on the cooler if the worts are run down at or above that temperature. It is while the wort is running over the refrigerator that such infection is most likely to occur, and as has been indicated above this can be almost entirely prevented by taking care that the refrigeration is carried out in an atmosphere of purified air.

From the above, it will be gathered that the operations of the brewer are greatly complicated and rendered much more difficult by the necessity of working under what may, for want of a better term, be called 'aseptic conditions.' Mere cleanliness is not sufficient, and the modern brewer has to be almost as constantly on his guard against the introduction of pathogenetic organisms as has the modern surgeon. In 1875 Pasteur, as the result of his epoch-making labours, wrote the two following sentences:

'Every unhealthy change in the quality of beer coincides with the development of micro-organisms foreign to brewers' yeast, properly so-called.'

'The absence of change in wort and beer coincides with the absence of foreign micro-organisms.'

These two short statements embodied a great truth, and virtually marked the transition from darkness to light, and from chaos to order. Somewhat extended and modified as the result of more recent research they are recognised as the foundation stone on which so much of modern brewery practice is built, and without which real success, if attained at all, would be largely a matter of chance. It has already been stated that in some breweries where the cooler has been abolished there is not even a collecting vessel, and the wort is pumped from the hop-back directly over the refrigerator. I do not think that this is altogether the best system, but it cannot be denied that in many cases it yields perfectly satisfactory results. A greater strain is, of course, thrown on the refrigerators which must have a larger cooling capacity than would otherwise have been necessary, that is to say, if the wort is to be cooled down in reasonable time. It is customary in some breweries to pass artificially cooled water through the lower tubes of the refrigerators, and thus to increase their cooling power. When there is refrigerating machinery the plan is a good one. While the wort is passing over the refrigerator in a thin layer it absorbs oxygen from the air, that is, of course, as soon as the temperature has fallen sufficiently low. This dissolved oxygen is of importance during the next stage of the process since it is necessary for the proper activity of the yeast.

CHAPTER VI

FERMENTATION

THE wort having been cooled to about 60° F. passes from the refrigerator to the vessel in which the process of fermentation is to be carried out. Here the requisite amount of yeast is added and the contents of the vessel are thoroughly mixed. Before proceeding farther, it will be necessary to devote some little space to the consideration of yeast as a living organism, and then to discuss at somewhat greater length the nature of the process of fermentation so far as it is at present known. If a little ordinary brewers' yeast be mixed with water and examined by means of the microscope, it will be found to consist of a number of bodies, some approaching the spherical in their contour, others being more or less ovoid (fig. 4)

Each of these is a self-contained organism,

consisting of a single cell. On examining these cells more closely, it will be seen that each is bounded by a cell wall which encloses the *protoplasm* and other cell-contents. In the younger cells the protoplasmic contents are clear and transparent, but as the cell

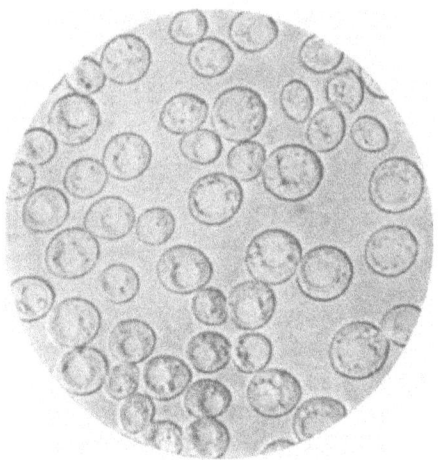

Fig. 4. ×750. *Saccharomyces cerevisiae*.

grows older, the protoplasm becomes more granular in character and one or more cavities known as 'vacuoles' may be observed. These vacuoles consist of the cell juice, which at certain stages of development is disseminated throughout the protoplasm but later tends to collect in one or more parts of the

vi] FERMENTATION 75

cell. These cells are very minute, having an average diameter of only $\frac{1}{120}$ millimetre ($\frac{1}{3000}$ inch) and an estimated volume of 0·0000004 cubic millimetre. It has been calculated that an ounce of pressed yeast would contain no fewer than 50,000 million cells. Minute as they are, however, each of these cells is

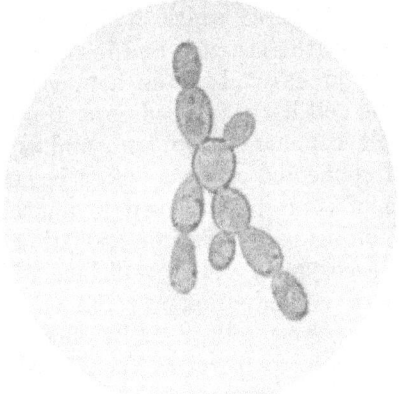

Fig. 5. × 660.   *Sacch. cerevisiae* budding.

the seat of vital processes of the greatest complexity, and may, without exaggeration be said to constitute a laboratory in which are carried out chemical changes which the most highly trained modern chemist is powerless to imitate. If a little young yeast is examined microscopically, it will be seen that many of the cells are not single, but have smaller cells attached to them, whilst in some cases chains of three, four or even more may be observed (fig. 5).

This is the chief mode of reproduction, namely by budding. The bud occurs first as a small protuberance on the surface of the cell. This quickly increases in size, until it has attained roughly the dimensions of the parent cell, after which it becomes detached, leading a separate existence, and in turn reproducing by the same process. It often happens that before the offspring cell has separated from the parent cell, it has itself commenced to bud, and so chains or clusters of connected cells may often be seen. Yeast is capable of reproducing itself in another manner than by budding, namely by the formation of internal spores or ascospores. The conditions which favour this mode of reproduction are the employment of young and vigorous cells, a moist surface, plenty of air and a suitable temperature (usually about 25° C.). Under these circumstances and at the end of about 24 hours, certain changes will be seen to be taking place in the protoplasmic contents of many of the cells. The protoplasm becomes at first more granular and then signs of segregation become visible, the contents of the cell separating into several ill-defined portions, usually from two to four, but in some species as many as eight. A little later these segregated portions of highly granular protoplasm become invested with a membrane and it can then be seen that the original cell contains sometimes one, but usually two or more well-defined spores (fig. 6).

During the formation and development of the

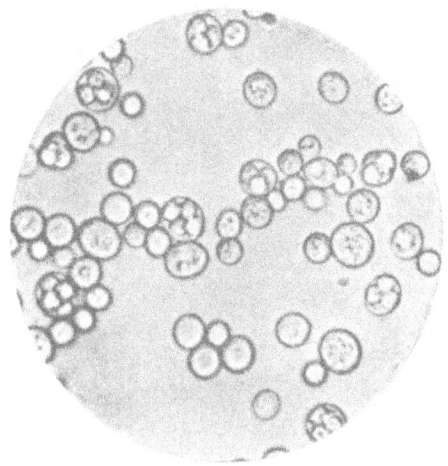

Fig. 6. ×750. *Sacch. cerevisiae* showing ascospores.

spores the parent-cell swells considerably, and in the end bursts, liberating the spores, each of which constitutes an individual yeast cell and is capable of reproducing in the ordinary way by budding. During recent years much study has been devoted to the precise mechanism of reproduction in the case of yeast. For long it has been known that every yeast

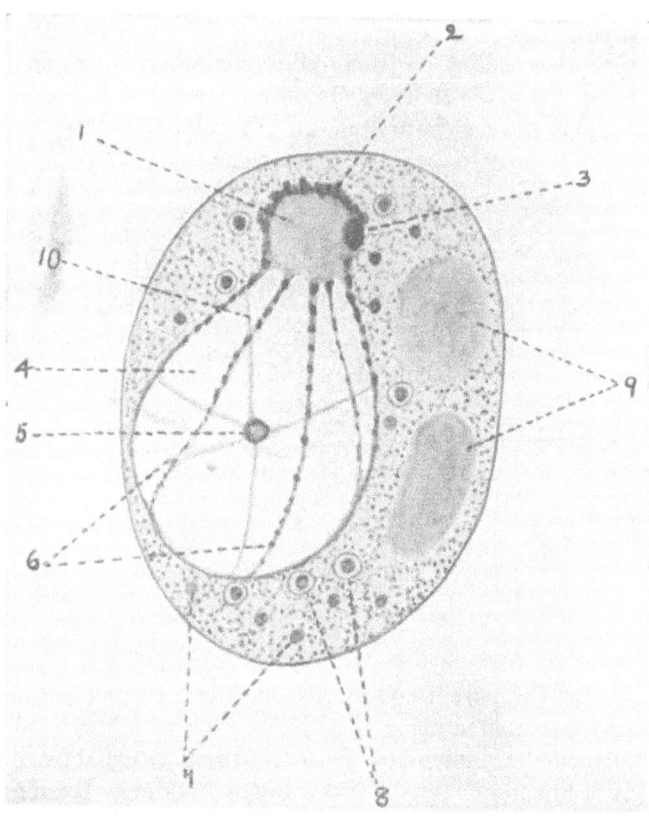

Fig. 7. 1. Nucleolus. 2. Peripheral layer of chromatin. 3. chromatin patch on one side of nucleolus. 4. Nuclear vacuole. 5. Central volutin granule in the vacuole. 6. Chromatin network. 7. Granules of fatty substance. 8. Volutin granules. 9. Glycogen vacuoles. 10. Delicate suspending threads for the central volutin granule. (After Wager.)

cell contained a nucleus but the true nature and function of this has only comparatively recently been made clear through the work of Janssens and Leblanc and especially of Wager and Peniston. The last mentioned authors have shown that every cell possesses a 'nucleus' and a nucleolus and they have adduced a good deal of evidence to prove that the former is identical with the main vacuole of the cell, and that the latter is a homogeneous body which is always found to be in close contact with the nuclear vacuole (fig. 7).

During budding, division of the nucleus takes place, accompanied by constriction into two approximately equal parts, one of which passes into the bud, whilst the other remains in the parent cell. During ascospore formation the vacuole disappears and only the nucleolus remains. This, however, divides into two by a process of constriction, and each of these two parts, again divides into two so that four nuclei are formed, each of which becomes the nucleus of a spore, and consequently of a new cell. Barker has called attention to a process of true conjugation in the case of a certain species to which he gave the name *Zygosaccharomyces*. Under certain conditions cells of this yeast formed buds which gradually developed into long beak-like processes. When the 'beaks' of two adjacent cells touched one another union took place, the tips of the beaks disappeared and a tubular connection was so established between the two cells, each of which then produced one or more spores. Interesting as these phenomena are to the biologist they do not closely concern the practical brewer, since under the ordinary conditions of the brewery yeast always reproduces by the process of budding. Unlike certain other uni-cellular organisms and the higher plants the yeast cell contains no chlorophyll, and is not therefore able to obtain its carbon by the decomposition of carbon-dioxide. In this respect it resembles the fungi, in which great natural family it is included. Owing to its ordinary mode of reproduction, it is classed among the *budding fungi*, and lastly its ability to form ascospores completes its claim to belong to the genus *Saccharomyces*, a group which comprises all the important organisms which produce alcoholic fermentation. Like all fungi, the yeast organism uses up oxygen and gives off carbon-dioxide, and it is for this reason that a supply of oxygen is necessary if the vigour of the yeast is to be maintained during fermentation. This point will be referred to again when dealing with the theories which have been put forward to explain the process of fermentation. A close investigation of the Saccharomycetes has revealed the fact that the genus includes a considerable number of species, many of these differing widely in their fermentative and other properties, and that certain of these species can be again subdivided into races or varieties. Even at a comparatively early period certain well-marked morphological differences were noticed as the result of microscopical examination. Thus some yeasts formed spherical or ovoid cells, whilst others were decidedly ellipsoidal, and others again elongated and sausage shaped. No great advance, however, could be made in differentiating between the various species until the late Professor E. C. Hansen in 1879 showed how it was possible to obtain almost any quantity of yeast by starting with *a single cell*. For this purpose a very dilute mixture of yeast and sterilized water is first made, and a little of this is inoculated into a quantity of melted wort gelatine,—that is to say, a moderately strong malt wort containing sufficient gelatine to cause it to solidify when cold. A drop of this solution is then examined by means of the microscope to ascertain whether it is sufficiently dilute in respect of yeast cells,—that is to say, whether the cells are well separated from one another and are so far apart that the colonies resulting from their development could not possibly meet. If such is the case a drop of the gelatine mixture is spread out in a thin layer on a microscope cover-glass, on which it solidifies. It is then placed, with the gelatine downwards, over a small glass cell containing a drop of water to keep the gelatine surface moist. Several yeast cells which are well separated from one another, are then picked out by microscopical examination and their position on the glass cover marked. The slide is then kept at a suitable temperature, and in a few days the development will have proceeded so far that the resulting colonies will be visible to the naked eye. When this is the case a very small piece of sterile platinum wire is dipped into any one of the colonies and then dropped into a suitable flask containing sterile wort. In the course of several days the wort will be found to be in a state of active fermentation and a sufficient quantity of yeast will have been formed for the 'pitching' of a still larger quantity of wort. Working in this way and pitching each time into vessels of larger size, it will be seen that practically any quantity of yeast can be prepared the whole of which has originated from a single cell. Pure cultures of the various Saccharomycetes having been thus prepared, it was found that morphological characters were frequently useless for the purpose of distinguishing between one species and another, for not only did many of these resemble one another somewhat closely in appearance, but the shape of any one species varied within wide limits, depending chiefly upon the conditions under which it had been grown. This method of obtaining pure cultures when used in conjunction with certain other methods of differentiation, such as the behaviour of the yeasts towards certain selected carbohydrates, and the

FERMENTATION

optimum temperatures for ascospore and film formation, has enabled zymotechnologists to isolate and describe many distinct species, of some of which numerous varieties are known. It should be said at once, however, that only a few of these are of industrial importance. For technical purposes the yeasts may be divided into two classes, the 'cultivated' and the 'wild' yeasts. The former includes brewers' yeast in all its varieties, that is to say, the yeast which has from the very earliest times been used for the production of alcoholic beverages, and has, in a sense, been cultivated for the purpose. This yeast represents so far as is known one species, namely *Saccharomyces cerevisiae*, although many races and varieties are known which differ considerably in certain respects, as for instance in the rapidity with which they bring about the fermentative change, the degree of attenuation (i.e. fermentation) which they can effect, and the flavour of the beer produced. Of the *Saccharomyces cerevisiae* there are two main types known respectively as 'top' and 'bottom' yeast. The former rises to the surface during fermentation, and is the yeast used in English breweries, whilst the latter sinks to the bottom of the fermenting tun, and is used in the production of lager beer as brewed on the Continent and elsewhere. The 'wild' yeasts are yeasts which occur wild in nature, and frequently having their habitat on the surface of ripe fruits, often find their way into the brewery. Some of these wild yeasts, (using the term in its widest sense) are capable of fulfilling useful functions in connection with cask fermentation, but others are highly undesirable. Although in his study of the various 'diseases' to which beer is subject Pasteur chiefly concentrated his attention on the bacteria, he did not altogether overlook the possibility that certain of the yeasts might be pathogenetic in character. It will be obvious, however, that no definite information in this connection could be obtained until Hansen had shown how to discriminate between the various species. It was then found that certain species of yeast were as much to be feared as many of the bacteria. Thus *Sacch. Pastorianus I* produces a nauseous bitter flavour and a disagreeable smell; *Sacch. Pastorianus III* and *Sacch. ellipsoideus II* persistent turbidity, *Sacch. anomalus* a pronounced fruity flavour, *Sacch. ilicis* a disagreeable bitter flavour, and *Sacch. foetidus*, stench. The following photomicrographs will give some idea of the microscopical appearance of a few of these 'wild' yeasts. *Sacch. ellipsoideus* (fig. 8.) *Sacch. Pastorianus* (fig. 9). *Sacch. anomalus* (film) (fig. 10) and *Sacch. apiculatus* (fig. 11). The last mentioned organism ought not properly speaking to be included among the Saccharomycetes, since it has never been observed to form endogenous spores, but its appearance is very characteristic, and it often finds its way into the cooling wort. The following may perhaps be regarded as the more usual ways in which infection with 'wild' yeasts takes place.

(1) Direct aerial infection, usually at the re-

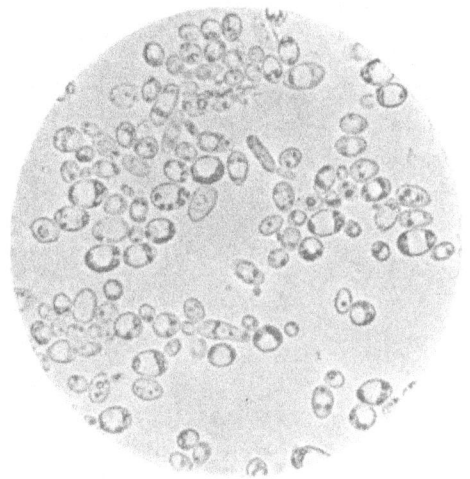

Fig. 8. ×750. *Sacch. ellipsoideus*.

frigerating stage, more rarely on the cooler or in the fermenting tun.

(2) Indirect aerial infection; that is to say, by dust which has accumulated on prominent internal surfaces of the cooler, refrigerator or fermenting-tun rooms, and which has become dislodged by wind.

(3) Surface infection, or infection due to 'nests' formed in the soft or old parts of fermenting tuns or yeast backs.

(4) Infection due to unusually impure pitching yeast, that is, the yeast used to start the fermentation.

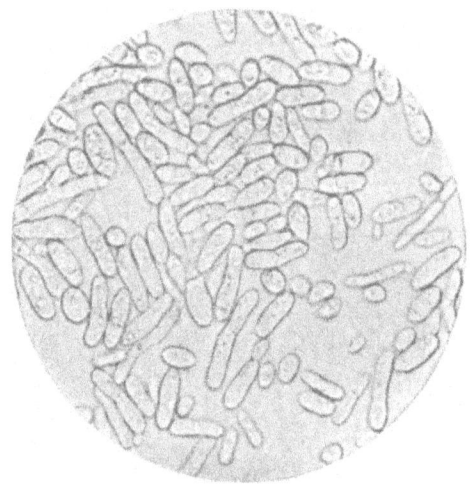

Fig. 9. ×750. *Sacch. Pastorianus.*

Owing to the tendency of culture yeasts to oust the wild species, especially under the conditions of English brewing, it will not often be found that serious trouble can be traced to this cause.

(5) Infection due to dry-hopping—that is the

introduction of wild yeasts with the hops added to the beer in cask.

How the more important sources of such infection may be guarded against has already been pointed out in the previous chapter.

Since even different races or varieties of the

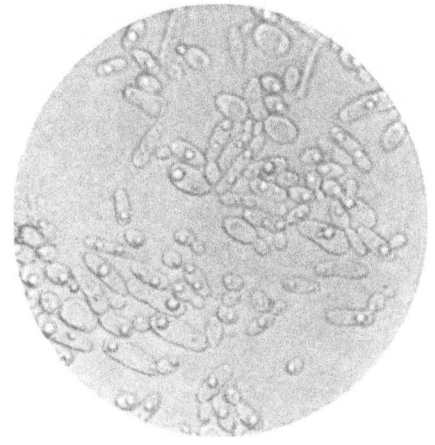

Fig. 10. × 750. *Sacch. anomalus* (film).

cultivated yeast *Sacch. cerevisiae* exhibit different properties, the introduction of the method of preparing pure cultures from a single cell naturally raised the question whether certain selected varieties (or even species) might not be successfully used in practice. Such pure cultures were first introduced by Hansen himself into certain Danish breweries with excellent results, and the method spread so rapidly on the Continent as almost to constitute a revolution in Continental brewing practice. For each brewery, experiments had first to be made in order

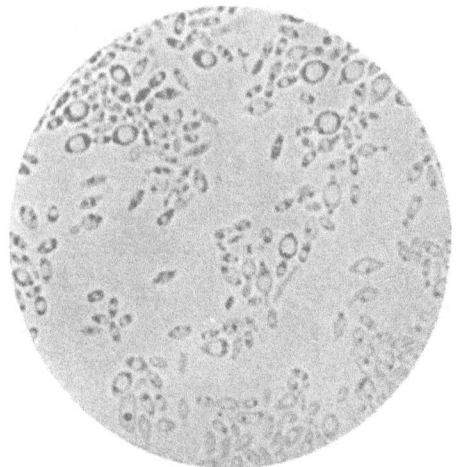

Fig. 11. × 750. *Sacch. apiculatus*.

to find out which of a number of varieties was the one best suited to the conditions obtaining in that brewery, and to the type of beer required, and then that yeast was cultivated in the necessary quantity. There can be no doubt that, as a general rule, the results have been very satisfactory, beers of greater stability, of more uniform character and of better flavour resulting. A good many attempts to introduce the use of selected single-cell yeast into English breweries have, however, met with much less success. One reason for this is that the conditions obtaining in most English breweries are such as to result in the production of a definite type of yeast, which gives the exact class of beer required, and which can without any special steps be kept practically pure, that is free from bacteria and other yeasts within the limits necessary for successful working. In the second place, there is a greater difference in character between the main fermentation and the secondary or cask fermentation in English high fermentation beers, than is the case in lager beers such as are brewed on the Continent; and it has not hitherto been found possible to obtain with a single-cell yeast the proper cask fermentation which is so important a feature in English brewing.

Having now considered yeast from the purely biological standpoint, it may be convenient to refer briefly to the question of its nutrition, that is to say, to the various substances which it requires for its healthy development, and then to consider how far these are naturally supplied in the brewers' wort. It has been pointed out above that yeast, like all organisms which are devoid of chlorophyll, cannot obtain its carbon from carbon-dioxide, and it is, therefore, necessary to supply it in some soluble and assimilable form. The various carbohydrates represent one such form, and it is from these that the carbon is to some extent, but not by any means entirely obtained, and the necessary vital energy indirectly derived. Of the protoplasm, nitrogen is the most significant and important constituent, and dried yeast contains about 8 per cent. or more of that element which is equivalent to 50 per cent. or more of protein matters. For the manufacture of its protoplasm, therefore, it is essential that sufficient nitrogen should be supplied to the yeast organism, and in the form in which it can be most readily absorbed and assimilated. Yeast is, in fact, able to satisfy its nitrogen requirements when presented with such simple forms as ammonium salts, but its development is more vigorous when the nitrogen is supplied to it in the form of amino compounds, amides and peptones,—that is to say when it is grown in solutions containing the products of the hydrolysis of proteins. The more complex proteins cannot apparently be utilized, since they are not diffusible and cannot therefore penetrate the cell wall. Another element necessary to the life-activity and well-being of the yeast organism, is phosphorus, which is present in the form of nucleo-proteins and to some extent as phosphate. Of the inorganic elements potassium, magnesium, and to a less extent calcium are indispensable as may be gathered from the following analysis of the ash of yeast:

Average composition of the ash of yeast (Lintner).

Potash (K_2O) with a little soda	33·49
Magnesia (MgO)	6·12
Lime (CaO)	5·47
Oxide of Iron	0·50
Phosphoric Acid (P_2O_5)	50·60
Sulphuric Acid (SO_3)	0·56
Silica	1·34
Matters undetermined	1·92
	100·00

The need for oxygen has already been referred to. Now it so happens that all these necessary nutritive substances are normally present in malt wort, which constitutes in fact an almost ideal liquid for the nourishment of yeast. There are the carbohydrates (sugars), the amides (e.g. asparagin), peptones and other diffusible and assimilable nitrogenous substances, phosphorus as phosphate, and sufficient quantities of the salts of potassium, magnesium and calcium. It does, however, happen occasionally, owing to the employment of malts of abnormal character, or to the use of larger proportions than usual of grain or sugar adjuncts, that the worts are deficient in the precise kinds of nitrogenous and other nutriment needed by the yeast for its proper development. In such cases the appearance of the yeast in the fermenting tun is such as to indicate that it is not receiving the food it requires, and the brewer usually endeavours to make good the deficiency by adding to the fermenting wort a quantity of so-called 'yeast food' containing the needful nutrient materials, and specially prepared for the purpose. Pressed yeast, that is yeast practically free from extraneous water, contains as a rule about 75 per cent. of water, present of course in the liquid protoplasm of the cells. The following analysis shows the average composition of *dry* yeast, and will suffice to give some idea of the proportions in which its more important proximate constituents are present:

Average composition of dry yeast.

Proteins and other nitrogenous substances ...	51·8
Yeast gum and other carbohydrate matter ...	29·5
Fat	1·0
Mineral matter	11·0
Matters undetermined, including some cellulose	6·7
	100·0

Having now considered the nature of the yeast as a living organism, and having dealt with its food requirements, we may pass to a consideration of the interesting and important phenomenon which it is its special function to excite, and which is known as *fermentation*. Few chapters in science are more fascinating than that dealing with the elucidation of the mechanism of the process of alcoholic fermentation, a process which was deliberately carried out thousands of years ago, and which could scarcely fail to attract the attention of the first man who left a bowl of honey exposed for sufficient time to the air. To deal at length with this question would be impossible within the limits assigned to this book, and as many of the earlier theories are now only of historical interest, I propose to go no farther back than the later views of Liebig. From the chemical point of view, fermentation consists essentially in the decomposition by means of yeast of a carbohydrate such as dextro-glucose into alcohol and carbon-dioxide, according to the equation:

$$C_6H_{12}O_6 = 2C_2H_6O + 2CO_2$$
$$\text{d-glucose} \quad \text{alcohol} \quad \text{carbon-dioxide.}$$

As a matter of fact the chemical reactions involved are by no means so simple as the above equation would indicate, since whilst 95 per cent. of the change takes place in accordance with that equation, about 5 per cent. of the sugar is broken up into glycerine, succinic acid and other products. A small proportion of the sugar is also utilized by the yeast for the formation of new cells. In 1839 Liebig put forward the view that a ferment (e.g. yeast) is a nitrogenous substance in a state of molecular instability, and that it was a peculiarity of such substances that they were able by mere contact to communicate this state to certain compounds such as the sugars, causing these to break down into simpler substances such as alcohol and carbon-dioxide. This theory of contact decomposition as the result of molecular vibration held the field for more than 30 years, but during the latter part of its life it was vigorously attacked by Pasteur, who was then engaged upon his epoch-making biochemical researches. The controversy was a long and vigorous one, and in the end Liebig so far modified his original views as to admit that the fermentation process was in some way connected with the life activity of the organism producing it, but he adhered so far to his earlier view as to hold that this life activity was not in itself the exciting cause, but was only necessary for the formation of some protein-like substance which actually brought about the decomposition. This theory which was put forward shortly before his death in 1873 is of special interest in connection with the view which is now universally held as the result of comparatively recent observations. To Pasteur belongs the credit of having been the first to prove definitely that fermentation was a physiological phenomenon, that is to say that it was intimately connected with the development and general vital activity of the yeast cell. He found that when a small quantity of yeast was introduced into a fermentable liquid saturated with air (oxygen) the oxygen was quickly absorbed and the yeast multiplied freely. He further found that when the whole of the free oxygen had been used up, fermentation proceeded vigorously, but was accompanied by comparatively little yeast reproduction. From experiments of this kind, made with several 'organised ferments' and under various conditions, he was led to the discovery that whilst certain organisms were only capable of living when supplied with plenty of free oxygen, others could survive the total deprivation of that element, and when so deprived appeared to function most actively as ferments. From this he was led to conclude that there was a very intimate connection between the phenomenon of fermentation and life in the absence of air. Organisms which required free oxygen he called 'aerobic' and those which could live without it 'anaerobic.' Yeast was an organism capable of living under both sets of conditions, but it was only as an anaerobe that it functioned at all actively as an alcoholic ferment. It may be of interest to quote Pasteur's own words:

'Fermentation by yeast, that is to say, by the type of ferments properly so-called is presented to us, in a word, as the direct consequences of the processes of nutrition, assimilation and life, when these are carried on without the agency of free oxygen. The heat required in the accomplishment of that work must have necessarily been borrowed from the decomposition of the fermentable matter, that is, from the saccharine substance, which like other unstable substances, liberates heat in undergoing decomposition. Fermentation by means of yeast appears, therefore, to be essentially connected with the property possessed by this minute cellular plant of performing its respiratory functions, somehow or other, with oxygen existing combined with sugar. Its fermentative power...varies considerably between two limits fixed, by the greatest and least possible access of free oxygen which the plant has in the process of nutrition. If we supply it with sufficient quantity of free oxygen for the necessities of its life, nutrition and respiratory combustions; in other words, if we cause it to live after the manner of a mould, properly so-called, it ceases to be a ferment....On the other hand, if we deprive the yeast of air entirely, or cause it to develop in a saccharine medium deprived of free oxygen, it will multiply just as if air were present, although with less activity, and under these circumstances its fermentative character will be most marked.' In a word, the yeast cell required oxygen for its vital activities, and if deprived of free oxygen would obtain that which it needed by breaking up the fermentable sugars, and utilising the oxygen which was supposed to be liberated in the process. This theory of 'life without free oxygen' was almost universally accepted as an explanation of the process of fermentation for a period of nearly 30 years, although it obviously did not supply a very clear insight into the mechanism of the change. It purported to tell us *why* the yeast cell decomposed the sugar, but not *how* it did it, and had any chemist been challenged during that period to state more precisely what occurred when the sugar molecule was broken up into alcohol, carbon-dioxide, and other substances, he would probably have referred it to some special vital force resident in and indissolubly connected with the living protoplasm. In 1892 Adrian J. Brown, in the course of an investigation on certain phenomena connected with the reproduction of yeast, obtained some results which rendered the 'life without free oxygen' theory quite untenable. By applying for the first time the haematimeter to the counting of the cells, he found that when yeast was introduced into a fermentable liquid such as wort, the cells increased to a definite number for a given volume, and then ceased to multiply. The limit of increase was within certain limits independent of the food-supply in the liquid, and also of the number of cells originally added, provided that number did not exceed the maximum number to which the cells would increase under normal conditions in the liquid in question. If on the other hand a number of cells exceeding the maximum were added, little or no multiplication occurred. Hence by working with a number of cells in excess of the maximum, multiplication could be avoided, and fermentation phenomena could be studied under conditions which practically eliminated that disturbing factor. Working in this way, Brown found that aerobic conditions did not inhibit fermentation, but that on the contrary a plentiful supply of free oxygen actually stimulated it, and so was forced to the conclusion that Pasteur's view could no longer be upheld. There can be no doubt that the yeast cell does require free oxygen for its development and for the purpose of carrying out its life activities, and although its fermentative vigour may appear for a time to be independent of that element, a fresh supply must be forthcoming, or the yeast will languish and die. In 1897 a further and very great advance was made in our knowledge of the nature of the fermentation process, for in that year E. Buchner showed that the liquid contents of the yeast cell, when added to a fermentable liquid, are able to excite fermentation without the presence of any cells at all. He ground up pressed yeast with quartz sand or kieselgühr so as to rupture as many of the cells as possible, and after adding a little water and wrapping in a cloth, subjected the mixture to a pressure of about 500 atmospheres (3 or 4 tons per square inch). In this way he obtained a clear, slightly opalescent liquid which when added to a solution of sugar very soon brought about its fermentation, precisely as if yeast itself had been used. At first it was objected that small particles of living protoplasm had passed through and that it was in these that the fermentative activity resided. This, however, was shown not to be the case, for not only could the expressed yeast juice be evaporated to dryness at a low temperature without losing its activity, but it was capable of bringing about fermentation in the presence of certain substances such as chloroform, and arsenites, which are protoplasm poisons, and are known to exert a powerfully inhibitory effect on the life of the cell. As a further proof that living protoplasm was not in question it was found possible to filter the juice through a Chamberland filter without destroying its activity. From these and many other experiments, Buchner concluded that fermentation was the result of the activity of an enzyme secreted by the yeast cell to which he gave the name 'zymase.' As in the case of other enzymes, the activity of zymase is very dependent on external conditions. Thus when the yeast juice is heated to a temperature of about 50°C. coagulation occurs and the filtered liquid has no longer any appreciable action on sugar solutions. In the same way its activity is very greatly reduced or altogether destroyed by certain substances such as prussic acid. The next advance in our knowledge of the mechanism of fermentation is due to Harden and his colleagues. Harden found that when he submitted yeast juice to filtration through a Chamberland filter impregnated with gelatine, he obtained a filtrate and also a residue on the filter. On experimenting with these two portions into which the yeast juice had been resolved, he found that neither was possessed of any power of bringing about fermentation. When, however, the two portions were mixed, their activity was restored. In this way Harden was led to recognise that the enzyme (the portion which remained on the filter) is powerless to produce fermentation unless in contact with the portion which had passed through, the active constituent of which he designated the 'co-enzyme.' The solution of the co-enzyme, the true chemical nature of which is still undetermined, retains its activity even after boiling. A further flood of light has been thrown on the nature of the fermentation process as the result of Harden's investigations on the effect of adding alkaline phosphates to sugar solutions in course of fermentation. Such addition was always followed by a rapid increase in the evolution of carbon-dioxide, and it was found that a definite relationship existed between the amount of phosphate added and the volume of carbon-dioxide liberated, a molecular proportion of the phosphate always resulting in the disengagement of a molecular proportion of carbon-dioxide. An attempt to ascertain precisely what happened to the added phosphate in these experiments resulted in the discovery of a new compound consisting of a hexose[1] sugar residue and phosphoric acid, apparently having the composition $C_6H_{10}O_4(H_2PO_4)_2$ and designated *hexose phosphate*. When the phosphate is added to a fermentable solution in the presence of zymase and the co-enzyme, it is supposed that one molecule of sugar (dextrose) breaks down into alcohol and carbon-dioxide, whilst a second molecule reacts with the phosphate to form the hexose phosphate, the two reactions proceeding in accordance with the following combined equation :

$$2C_6H_{12}O_6 + 2Na_2HPO_4$$
$$= 2CO_2 + 2C_2H_6O + 2H_2O + C_6H_{10}O_4(Na_2PO_4)_2.$$

In practice it is well known that worts containing only a limited amount of phosphate, but relatively very large proportions of fermentable sugar, are capable of undergoing complete fermentation, and it is clear that according to the above view, this could not happen unless the phosphate were in some way or other regenerated. Harden considers that the yeast cell contains an enzyme to which he has given the name 'hexosephosphatase,' the function of which is to effect the hydrolysis of the hexosephosphate in the following manner:

$$C_6H_{10}O_4(Na_2PO_4)_2 + 2H_2O = C_6H_{12}O_6 + 2Na_2HPO_4.$$

[1] A hexose is a sugar containing six carbon atoms in its molecule, such as dextrose.

Summarising the above statements the following may be regarded as the most recent view of the fermentation change.

The enzyme zymase and its co-enzyme together act on the sugar (hexose) in the presence of the phosphate, in such a way that one half of the sugar is decomposed into alcohol and carbon-dioxide, whilst the other half unites with the phosphate to form the hexosephosphate above referred to. The phosphate is thus rendered temporarily inoperative, but is liberated by the action of the enzyme 'hexosephosphatase,' which reproduces the sugar and phosphate, and so the cycle of change is ready to be repeated. This theory at least explains all the facts at present ascertained, but there can be little doubt that very much work remains to be done before we can feel satisfied that we know exactly what happens when a molecule of sugar is decomposed at the instance of the yeast cell. As in all cases where living organisms are concerned, the process is one of very considerable complexity, and compared with the apparent ease and rapidity with which the wonderful changes, summed up in the term 'fermentation,' are accomplished by the yeast organisms, the triumphs of the modern organic chemist pale into insignificance. In the words of Professor Meldola:

'When we can transform sugar into alcohol in the laboratory at ordinary temperatures by the action of a synthesised nitrogenous organic compound: when we can convert glucose into citric acid in the same way that Citromyces can effect this transformation ...then may the chemist proclaim with confidence that there is no longer any mystery in vital chemistry.'

Attention has already been called to the fact that alcohol and carbon-dioxide, although by far the most important, are not the only products of alcoholic fermentation, and it has been stated that glycerine, succinic acid and various higher alcohols are formed at the same time. Recent investigations have thrown a great deal of light on the formation of these by-products, and it has been found that some of them at least are not formed from the sugar at all. Thus Ehrlich has shown that the higher alcohols (so-called fusel oil) result from the action of yeast on the amino-acids, some of which are always present in malt-wort, —leucine (α-amino-iso-caproic acid), for example, yielding amylic alcohol. In this reaction ammonia (which is assimilated by the yeast) and carbon-dioxide are liberated and an alcohol is formed containing one atom of carbon less than the original amino-acid. This reaction is probably of considerable importance from the point of view of the nutrition of the yeast, as it appears to indicate the means by which the yeast supplies itself with nitrogen in a readily assimilable form. The succinic acid which is a constant product of fermentation to the extent of about 0·5 per cent., appears to be derived from the dicarboxylic amino-acid, glutamic acid.

It has already been stated that some of the 'wild' yeasts produce bitter products, and certain species (e.g. *Sacch. anomalus*) give rise to the production of pleasant smelling ethereal substances such as are present in some fruit juices. It will, therefore, be understood that these by-products are of more than scientific interest, since the flavour of a fermented beverage is not unfrequently dominated by them, and its commercial value either greatly increased or greatly reduced according to their nature. In addition to zymase and the hexosephosphatase, the yeast cell contains a number of other enzymes, each of which has its specific function and all of which are of considerable importance in connection with the nutrition and life-activities of the organism. Incidentally some of these are, as will be seen, of considerable technical importance. So far as is known the hexoses alone (and of these only four, viz. d-glucose, d-mannose, d-galactose and d-fructose) are directly fermentable by yeast, and before the fermentation of other sugars can take place it is therefore necessary that they should be converted into one or other of these hexoses. This is in all cases effected by enzymes. Thus cane-sugar is not directly fermentable but has in the first instance to be converted by the enzyme invertase into a mixture of d-glucose (dextrose) and d-fructose (laevulose) in accordance with the following equation:

$$C_{12}H_{22}O_{11} + H_2O = C_6H_{12}O_6 + C_6H_{12}O_6.$$
$$\text{(cane-sugar)} \qquad\qquad \text{(dextrose)} \quad \text{(laevulose)}$$

Maltose has the same empirical formula as cane-sugar and is, prior to fermentation, converted by the enzyme maltase into two molecules of dextrose:

$$C_{12}H_{22}O_{11} + H_2O = 2C_6H_{12}O_6.$$
$$\text{(maltose)} \qquad\qquad \text{(dextrose)}$$

Milk-sugar (lactose) again, is not directly fermentable but has first to be resolved by the enzyme lactase into two other hexose sugars, dextrose and galactose:

$$C_{12}H_{22}O_{11} + H_2O = C_6H_{12}O_6 + C_6H_{12}O_6.$$
$$\text{(lactose)} \qquad\qquad \text{(dextrose)} \quad \text{(galactose)}$$

In this connection a point of considerable interest may just be mentioned in passing, namely that the enzymes exhibit a remarkable discrimination between certain carbohydrates which otherwise resemble one another very closely indeed. Whilst d-glucose, d-mannose, d-fructose, and d-galactose are fermentable, the optical isomerides of these carbohydrates,—that is to say, forms which differ only in respect of their action on polarised light,—are unfermentable. In order that a given sugar (other than the four above mentioned) may be fermented by yeast, it is essential that the yeast in question should contain the enzyme necessary for its conversion into one or other of the above hexoses. Now yeasts of different species do not all contain the same enzymes and it therefore happens that a certain yeast may be capable of fermenting one carbohydrate and incapable of fermenting another. Of the enzymes, invertase appears to be the most widely distributed among the Saccharomycetes, and consequently the great majority of yeasts are capable of bringing about the fermentation of cane-sugar. On the other hand lactase occurs in only a comparatively small number of species and consequently a great many yeasts, including the ordinary culture yeasts (*Saccharomyces cerevisiae*) are incapable of fermenting milk-sugar. The following table may be of interest as showing at a glance the behaviour of certain of the yeast species towards several of the more commonly occurring sugars. The sign + indicates that the yeast in question is capable, and the sign 0 that it is incapable of bringing about fermentation:

Yeast	Dextrose	Laevulose	Mannose	Galactose	Maltose	Cane-sugar	Milk-sugar
Sacch. cerevisiae	+	+	+	+	+	+	0
Sacch. cerevisiae Carlsberg	+	+	+	+	+	+	0
Sacch. Pastorianus	+	+	+	+	+	+	0
Sacch. ellipsoideus	+	+	+	+	+	+	0
Sacch. Marxianus	+	+	+	+	0	+	0
Sacch. exiguus	+	+	0	+	0	+	0
Sacch. Ludwigii	+	+	+	0	0	+	0
Sacch. anomalus	+	+	+	0	0	+	0
Sacch. fragilis	+	+	+	+	0	+	+
Kefir	+	+	+	0	0	+	+

It may be mentioned that the secretion of any particular enzyme is, so far as is known, a very constant attribute of a given species, and that it has not been found possible by varying the nature of the food supply or the general environment of a given species of yeast to cause it to secrete other enzymes than those normally present. This selective behaviour of the various species of yeast towards the different carbohydrates constitutes, in fact, a very valuable method of differentiation and identification. In addition to these sucro-clastic, i.e. sugar-splitting enzymes, the yeast cell contains one or more proteolytic enzymes,—that is enzymes capable of acting upon protein matters and of converting them by hydrolytic changes into simpler substances. Since the enzymes themselves are, chemically speaking, closely allied to the proteins, it might be thought that a proteolytic enzyme would attack the other enzymes and so bring about their destruction. As a matter of fact, such is the case, and under certain conditions zymase is attacked by the proteolytic enzyme of the yeast cell with the consequent destruction of its activity. There is a good deal of evidence, however, that in the living cell there exists a substance, whose special function it is to protect the enzymes from this action. There can be very little doubt that the fermentative and other changes above referred to take place *within* the yeast cell, and that the alcohol and carbon-dioxide having been formed in that wonderful little laboratory, then diffuse through the cell membrane into the surrounding liquid.

Having given a necessarily brief outline of the more important phenomena of fermentation both in their biological and chemical aspects, we may now return to a consideration of the process from the more purely technological side. It will be remembered that to the cooled wort collected in the fermenting vessel, the necessary amount of yeast is added, and fermentation allowed to proceed (page 73). The fermenting tun in its simplest form consists of a square or round vessel, usually constructed of some suitable wood, but often lined with copper, or occasionally at the present time, with aluminium. This vessel is fitted with a coil through which cold water may be circulated for the purpose of controlling the temperature of the fermenting wort, and an arrangement known technically as a 'parachute,' which is capable of adjustment in a vertical direction, and which serves for the removal of the yeast formed during the process and its transference to the 'yeast-backs' or store vessels below. In fig. 12 (a) is the attemperating coil, (b) the parachute, from the pipe (c) of which the yeast drops into a vessel on the floor beneath, and (d) a movable skimming arrangement for collecting and guiding the yeast into the opening of the parachute.

FERMENTATION

In the simplest system, known as the skimming system, the fermentation commences and ends in the one vessel, the beer being racked either directly from this vessel into the trade casks, or transferred to a

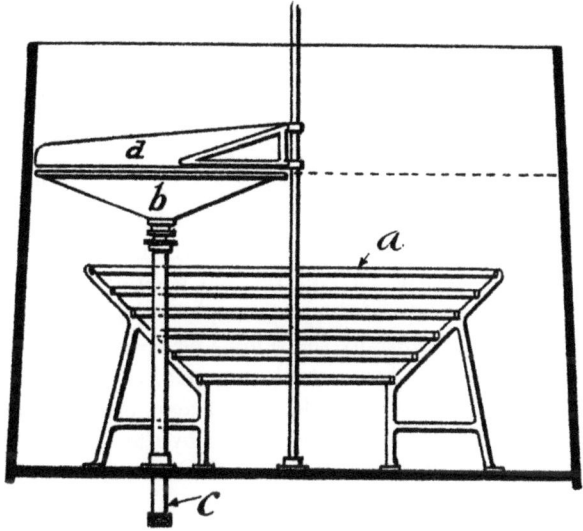

Fig. 12. Fermenting Tun.

racking vessel for purposes of convenience. In the latter, little or no fermentation occurs and the beer is rarely allowed to remain in it for more than 24 hours. One advantage of this vessel is that it enables the brewer when necessary, to add sugar solution, or 'priming' as it is called, in bulk instead of to the individual casks. In other systems of fermentation as practised in this country, the first part of the fermentation only is allowed to take place in the original fermenting tun, after which the partially fermented wort, is run into other vessels for the completion of the process, these vessels being either rectangular wooden vessels known as 'dropping squares,' or connected casks of special construction known as 'unions.' The latter system is typical of the working adopted in the breweries of Burton-on-Trent, and the vessels in question are usually known as 'Burton Unions.' Other systems are in vogue, some of which are peculiar to certain parts of the country, but whatever may be the precise nature of the plant employed, the essential process is substantially the same, and consists in the conversion of the fermentable sugars of the wort into alcohol and carbon-dioxide in the manner described above. The amount of yeast used by the brewer for the purpose of bringing about fermentation varies somewhat according to the strength of the wort and other circumstances, but may be given roughly as from one to three pounds of fairly solid yeast per barrel of 36 gallons. After the addition of the yeast, fermentation commences almost immediately, and at the end of a few hours the liquid in the tun will be found to be covered with a frothy or creamy layer. This rapidly increases in volume and a little later forms a fairly thick covering of very irregular surface, which is known as the 'rocky head.' This in turn undergoes a change in appearance, becoming more compact in character and presenting a more even surface. At this stage the progress of fermentation is rendered evident by the fact that the carbon-dioxide rising to the surface of the liquid, passes through this yeasty layer forming large bladdery bubbles which quickly burst and are immediately replaced by others. To any one watching the contents of such a tun the whole of the yeasty 'head' or covering will be found to be in a continual state of motion, owing to the escape of the carbon-dioxide gas. At this stage the newly formed yeast is skimmed off for the first time, or the partially fermented wort is transferred to the dropping square or union in which the fermentation is allowed to complete itself. During the progress of the fermentation the temperature of the fermenting wort rises owing to the vital activities of the yeast, and has for several reasons to be held in check. From 60° F. at which it is customary to add the yeast, the temperature rises to about 70° F. or 75° F., an increase beyond these temperatures being prevented by the passage of cold water through the attemperating coils above mentioned. It is not within the scope of this book to describe in detail the processes of fermentation as practised in the brewery, and it may therefore be assumed that the fermentation of the wort is completed and the liquid (now no longer wort but beer) is ready to be transferred to the trade casks or to be otherwise dealt with.

From all that has been said above, it will be clear that the essential feature of the fermentation process is the conversion of maltose into alcohol and carbon-dioxide through the agency of the yeast organism. As stated on page 105 the maltose has first to undergo conversion by the enzyme maltase into dextrose, and it is in reality this sugar which is broken up and furnishes the alcohol and carbon-dioxide. As fermentation proceeds the sugar is of course replaced by alcohol, and the carbon-dioxide simultaneously produced is liberated into the air, so that the specific gravity of the fermenting wort decreases continuously until the process is complete. This reduction of density is technically known as the 'attenuation' and as has been pointed out (page 37) is, within certain limits, capable of being controlled by the brewer during the mashing process. During fermentation the whole of the free maltose disappears, and it is also probable that some of the lower type malto-dextrins (page 35) are degraded to maltose and fermented by the ordinary so-called primary yeast (*Saccharomyces cerevisiae*). At racking, therefore, the beer will contain the following substances:

FERMENTATION

1. Alcohol
2. Carbon-dioxide
3. A little glycerine
4. Traces of higher alcohols, succinic acid, etc.

} Products formed as the result of the fermentation process.

5. Malto-dextrins of various kinds
6. The so-called stable dextrin
7. Proteins and other nitrogenous substances not utilised by the yeast
8. A little mineral matter
9. Bitters and other substances derived from the hops

} Unfermented constituents of the original hopped wort.

The above does not pretend to be a complete list, for beer contains many substances in minute proportions, but it is intended merely to show at a glance the more important constituents, and how these are roughly to be divided between the products of fermentation, and those present in the original wort. Analyses of a few representative types of beer will be found in the Appendix, page 126.

At the moment of racking, the primary fermentation has spent itself, that is to say, the whole of the readily fermentable carbohydrate matter has been decomposed and replaced by its equivalent of alcohol. A reference to the above table will show, however, that it still contains a certain amount of carbohydrate matter, which, as has been pointed out on page 33, is necessary for the cask or 'secondary' fermentation. Without this, the beer would rapidly become flat and undrinkable, and the brewer endeavours by every means at his disposal, to ensure that the proportion and the character of these residual carbohydrate matters (malto-dextrins) shall be properly suited to the class of beer he is brewing, that is to say, shall be such that the beer comes rapidly into condition on the one hand, or undergoes a slow and protracted cask-fermentation, on the other. Whilst the higher type malto-dextrins are not appreciably attacked by ordinary yeast (*Sacch. cerevisiae*) they are reduced to free maltose by the hydrolytic action of certain of the 'secondary' yeasts (e.g. *Sacch. Pastorianus*) and it is these yeasts which are largely responsible for the true secondary fermentation of stock beers. These yeasts are almost invariably present in very small proportions in the ordinary pitching yeast (i.e. the yeast added to the fermenting tun to start the fermentation) and are often conveyed to the wort from the air during the process of cooling. In any case they are rarely if ever absent from the beer at racking, although during the main fermentation they have been kept severely in check owing to the greater activity and enormous numerical preponderance of cells of the culture yeast. At the end of the primary fermentation, however, the latter begin to find the environment unsuitable, and the secondary yeasts, which are better able to attack the malto-dextrins, come into play and rapidly increase in numbers. In the case of mild ales and certain other beers which undergo practically no storage, but are drunk within a few days of being racked into cask, these secondary yeasts are not needed, for the fermentation which takes place in the cask during those few days is virtually a continuation of the primary fermentation, and not a true 'conditioning' in the strict sense of the term. It is for this reason that 'pure yeast,' that is, yeast of a single species, has in such cases been found to give good results in English breweries, whilst for stock ales it has invariably proved unsuitable. Owing to the presence of the alcohol, the carbon-dioxide and the hop-resin, all of which are bactericidal in varying degrees, beer is not at all a favourable medium for the growth of the great majority of bacteria. There are, however, some which are capable of thriving in it, and these may very easily give rise to disastrous results. Thus certain bacteria, prominent among which is the *Bact. aceti*, promote the oxidation of some of the alcohol to acetic acid, and so cause the beer to become sour. In vinegar making, this organism is intentionally used for the purpose of converting the alcoholic wash into acetic acid, and, indeed, vinegar could be made from beer in this manner. Certain other bacteria produce traces of butyric and other strong-smelling fatty acids, the presence of which, of course, renders the beer thus attacked quite undrinkable. Other bacteria, again, produce the condition known as 'ropiness,' in which the beer has an oily appearance when poured from a vessel, or, in extreme cases, may be almost jelly-like in consistency. In practice it is impossible to brew beer under such conditions as *entirely* to prevent the introduction of bacteria, and all that the brewer can do is, by paying the most unremitting attention to the cleanliness of his vessels and plant, and by excluding all but filtered air from the wort during certain portions of the process, to reduce the infection to a practically safe minimum. Given a sufficiency of the preservative constituents of the hop, and a constant development of carbon-dioxide (conditioning), beers brewed under such conditions will keep sound for many months, and in the case of strong beers, where the larger percentage of alcohol helps, for years.

CHAPTER VII

DISTRIBUTION, INCLUDING BOTTLING

Speaking broadly beer is supplied by the brewer either in cask (draught beer) or in bottle. At the present time the staple beer of this country is the so-called mild ale, which is intended for rapid consumption. This beer is supplied to the wholesale trade in cask and is usually consumed within a few days of leaving the brewery. Whilst most well-brewed beer will become spontaneously bright if kept for a sufficiently long time under proper conditions, this beer is required to be brilliant within a very short time of racking into cask. To effect this it is customary to add 'finings.' Finings consist of a viscous 'solution' of isinglass in dilute acid, usually sulphurous or tartaric or both. The isinglass, which is not really dissolved by the acid, but is only in a state bordering on solution, immediately coagulates, when the finings are added to the beer, and in the process entraps as it were, all the suspended particles, carrying them either to the top or to the bottom and retaining them in the coagulum formed. One of the most striking changes which has taken place in the brewing industry during recent years is the enormous development of the bottle trade, largely at the expense of small casks. Originally all bottled beer was conditioned in bottle, that is to say, it was allowed to mature for a certain time in cask and was then transferred to bottles, which were stored for a sufficiently long time to enable the 'secondary' or in this case 'bottle' fermentation to take place. Slow fermentation taking place under these conditions, that is to say, in the virtual absence of oxygen and under pressure, resulted in the production of a type of beer which could not be produced in any other way, and which was, and still is, highly appreciated by the connoisseur. The system is regarded by experts as in many respects the best, and it is still adopted in the case of the finest pale ales, export beers and other beers of the highest quality. For a long time the bottle fermentation was thought to be due entirely to secondary yeasts, such as those which are chiefly responsible for the cask conditioning of draught beers, but during recent years it has been shown (first by Claussen) that certain organisms belonging to the group of Torulae are in reality the active agents. The Torulae are closely allied to the true Saccharomycetes, from which they differ chiefly in their inability to form ascospores. As compared with the yeasts (from which they are often indistinguishable in their miscroscopical appearance) these organisms (fig. 13) have usually fermentative properties of a low order, and many of them are incapable of fermenting the ordinary sugars, to say nothing of the comparatively stable residual carbohydrates of beer. Some, however, have this power, and it is highly probable that the characteristic flavour of certain bottled beers is to some extent the result of their activity. At the same time there is abundant evidence that the ordinary secondary yeasts are also operative.

One objection to this system will be obvious, namely that since the conditioning of the beer is due to fermentation occurring within the bottle,

VII] DISTRIBUTION 119

there must be a corresponding formation of yeast, that is to say, of sediment. About 30 years ago the practice of 'carbonation' as applied to bottled beer made its appearance, and was soon found to meet a

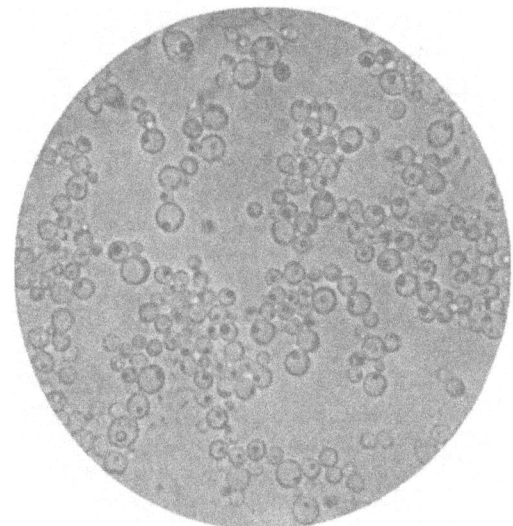

Fig. 13. ×750. Torula.

widely spread demand. In this process the beer was bottled in a bright condition and was then charged artificially with carbon-dioxide, as in the preparation of the ordinary mineral waters. From the brewer's point of view it had the advantage of enabling him to do a much quicker trade and to increase his turnover, whilst the consumer was glad to get a bottle of beer which was always charged with gas, and which being practically free from sediment, involved no waste. This process which is still widely adopted, and which has certain well-defined merits, is not technologically a very perfect one, for not only is it impossible to prevent effectually the occurrence of fermentation and consequently of sediment if the beer is not quickly consumed, but there is from time to time a certain amount of difficulty due to the separation of certain protein matters of slight solubility. This is often a source of very considerable difficulty to the brewer, in draught as well as in bottled beers. Certain of the proteins, which cannot be assimilated by the yeast, and which consequently pass through into the finished beer are frequently just on the borderland of true solution, and so a very slight alteration in the character of the solvent such as the formation of a little more alcohol, or the presence of a little more gas, causes them to come partly out of solution. In this state they cause the beer to present a hazy appearance, which is very difficult to deal with and which renders it for all practical purposes, unsaleable. Even the most perfect filtration could not entirely eliminate these defects from the simple carbonation process, and so brewers were quite prepared to welcome the 'chilling' system.

The principle underlying this is a very simple one. The solubility of the more complex protein matters dissolved in the beer diminishes with reduction of temperature, and if the beer is cooled sufficiently, they will come more or less completely out of solution. If then the beer is submitted to filtration at this low temperature, these matters will be removed, and the beer will subsequently remain brilliant for a considerable time, even when exposed to the most adverse conditions. In some cases the beer is cooled slowly in bulk in refrigerating chambers, whilst in others it is cooled rapidly in cylinders, but in both cases it is forced at a temperature but little above that of the freezing point of water through special filters before passing to the bottling machine. When the beer is cooled slowly in bulk there is a very perfect separation not only of the protein matters, but also of suspended yeast cells, and such beer will keep brilliant in bottle for a considerable time, provided that a thoroughly efficient filter is employed. Beer which has been *quickly* chilled is often less satisfactory in this respect. Unfortunately the lowering of temperature has the effect of bringing out of solution, not only the undesirable proteins but also the hop resin, and beers which have been submitted to this process frequently possess too little hop flavour, a defect which has not up to the present been entirely remedied.

Many modifications of these systems exist, but as the principles underlying them are practically the same, it will not serve any useful purpose to refer to them in detail. A certain amount of beer is sold in quart bottles or flagons, but as this is virtually draught beer, sold in bottles instead of in small casks, to suit the consumer's convenience, it is not necessary to refer to it further. It will be seen then that bottled beers may be roughly divided into three classes (*a*) beers matured in cask and allowed to condition in the bottles, (*b*) beers filtered and artificially gassed in bottle, and (*c*) beers chilled, filtered at a low temperature and bottled with or without the use of extraneous gas. That the beers of the first class are by far the best is generally admitted, and few beer drinkers would be found to deny the statement that in the other classes of bottled beer brilliancy is usually secured at the expense of character and flavour, although, of course, there are great differences in these respects in the products of different firms.

In concluding this Chapter, attention may be called to the utilization of two very important waste products of the brewery, namely carbon-dioxide gas and yeast. Taking the fundamental fermentation equation

$$C_6H_{12}O_6 = 2C_2H_6O + 2CO_2$$

it will be seen that 180 parts by weight of dextrose

yield no less than 88 parts by weight of carbon-dioxide, or for every ton of sugar fermented nearly half a ton of the gas is liberated. In many of the larger breweries this gas is collected from the fermenting tuns, purified by passing it through scrubbers and towers containing certain purifying agents, and then condensed, cooled and liquefied. The liquid carbon-dioxide is then transferred to stout cylinders, from which it is used by the brewer himself, or in which it is sold to other brewers or to mineral water manufacturers for the gassing of their products. Yeast is another by-product of the brewery to the utilization of which a very large amount of attention has been devoted during recent years. Under ordinary circumstances a reproduction equivalent to about five times the weight of the yeast used for pitching occurs, that is to say, for every one pound of fairly solid yeast added to the tun at the commencement of fermentation about six pounds of equally solid yeast will be recovered. Assuming 30 million barrels of beer to be brewed annually in the United Kingdom, and assuming further that five pounds of yeast will be obtained from each barrel fermented, it will be seen that the whole annual output of yeast for the United Kingdom alone would amount to no less than 67,000 tons. Of this, which is undoubtedly a very low estimate, a proportion is of course needed by the brewer for starting fermentations, but the great bulk is not so required, and is available for other purposes.

Of the various uses which have been suggested for the product, by far the most interesting and successful is its conversion into a food-product. It will be remembered that the yeast cell consists largely of protoplasm, the chemical basis of all living cells, and it is clear that it should possess nutritive properties of a high order. By a patented process certain changes are effected in the protein constituents, and as the result a product is obtained which resembles extract of meat so closely that it is almost impossible to distinguish between them either by taste or smell. In chemical composition, moreover, they are very similar indeed, the only difference so far as is known, being that meat extracts contain the bases creatine and creatinine, whereas yeast extracts do not. As a feeding-stuff for cattle, dried yeast is also being somewhat largely used.

APPENDIX

Table showing the number of bulk barrels of beer produced during the year ending 30th September 1911, and the quantities of malt, maize and rice, sugar and hops used during the same period:

Bulk barrels of beer	35,949,478
Malt	51,670,357 bushels
Rice and Maize preparations... ...	1,336,686 cwts.
Sugars of various kinds	3,011,201 cwts.
Hops	64,316,108 lbs.

Table showing the number of bulk barrels (that is, barrels containing 36 gallons) of beer produced in the United Kingdom during the years 1902–1910 inclusive, and the average specific (original) gravity for those years:

Year ended 31st March	Bulk Barrels	Average specific gravity
1902	36,887,260	1053·47°
1903	37,153,978	1053·42°
1904	36,329,813	1053·43°
1905	35,415,523	1053·24°
1906	35,066,094	1053·33°
1907	35,406,797	1053·31°
1908	35,359,024	1053·44°
1909	34,376,352	1053·26°
1910	34,299,914	1053·20°

Table showing the average composition of various kinds of beer.

	Extractive Matters	Proteins	Acidity (as lactic acid)	Ash	Absolute Alcohol (by weight)	Proof Spirit
London Mild Ale	5·15	0·33	0·12	0·28	3·65	8·10
Burton Pale Ale	4·32	0·50	0·234	0·31	5·58	12·16
English Strong Ale	7·59	0·91	0·162	—	8·75	19·00
English Light Bottled Ale	3·96	0·32	0·09	0·27	3·98	8·86
Dublin Stout	7·02	0·75	0·24	0·39	5·64	12·28
London Porter	5·52	0·37	0·13	0·37	4·37	9·54
American Ale	6·02	0·57	0·094	0·33	5·25	11·50
Export Pilsener	5·70	0·42	0·19	0·20	4·20	9·20
Export Munich	6·45	0·56	0·14	0·186	4·00	8·90

The following analysis shows in greater detail, the composition of a sample of Burton Pale Ale of medium gravity:

Total Extractive Matters	5·20 %
Combined Maltose	1·28 %
Combined Dextrin	0·38 %
Uncombined Dextrin	2·09 %
Fixed Acidity (as lactic acid)	0·04 %
Total Acidity (as acetic acid)	0·10 %
Proteins	0·48 %
Ash	0·31 %
Absolute Alcohol (by weight)	5·17 %
Proof Spirit	11·28 %
Phosphoric Acid	0·05 %

BIBLIOGRAPHY

HISTORICAL

'Curiosities of Ale and Beer.' JOHN BICKERDYKE. 1890.
'Drinks of the World.' J. MEW and J. ASHTON. 1892.
'In praise of Ale.' W. T. MARCHANT. 1888.
'Beer of the Bible.' JAMES DEATH. 1887.
'Origin and History of Beer and Brewing.' J. P. ARNOLD. Chicago 1911.
'History of Drink.' J. SAMUELSON. 1878.

SCIENTIFIC AND GENERAL

'A text book of the Science of Brewing.' E. R. MORITZ and G. H. MORRIS. 1891.
'The principles and practice of Brewing.' W. J. SYKES and A. R. LING. 1907.
'Practical Brewing.' E. R. SOUTHBY. 1895.
'A Handybook for Brewers.' H. E. WRIGHT. 1897.
'American Handybook of the Brewing, Malting and Auxiliary Trades.' ROBERT WAHL and MAX HENIUS. 1901.
'Laboratory Studies for brewing Students.' ADRIAN J. BROWN. 1904.
'A Manual of Brewing.' E. GRANT HOOPER. 1891.
'The Brewing Industry.' JULIAN L. BAKER. 1905.
'A Manual of Alcoholic Fermentation.' C. G. MATTHEWS. 1902.

'Laboratory Text Book for Brewers.' LAWRENCE BRIANT. 1898.
'The Hop and its Constituents.' Edited by A. CHASTON CHAPMAN. 1905.
'Hops.' E. GROSS. 1900.
'Micro-organisms and Fermentation.' ALFRED JÖRGENSEN. (Translated by S. H. DAVIES.) 1911.
'Fermentation Organisms.' A. KLÖCKER. (Translated by G. E. ALLAN and J. H. MILLAR.) 1903.
'Technical Mycology.' FRANZ LAFAR. (Translated by C. T. C. SALTER.) 1910.
'The Soluble Ferments and Fermentation.' J. REYNOLDS GREEN. 1901.
'Enzymes and their Applications.' J. EFFRONT. (Translated by S. C. PRESCOTT.) Vol. I. 1904.
'Ferments and their Actions.' C. OPPENHEIMER. (Translated by C. A. MITCHELL.) 1901.
'Études sur la Bière.' M. L. PASTEUR. 1876.
'An Atlas of the Saccharomycetes.' A. CHASTON CHAPMAN and F. G. S. BAKER. 1906.

The above list is not intended to be in any way exhaustive. It includes, as will be seen, only books dealing with the history of beer and other fermented beverages, and such standard works as might be found useful to anyone desirous of obtaining fuller information in regard to the scientific principles underlying the brewing and malting processes. With one exception, moreover, it includes only books which have been either published in, or translated into, the English language.

INDEX

Ale, origin of the word, 2
Amyloins, 31

Bacteria, infection with, 69, 115
Barleycorn, changes produced in during malting, 12–14, 17
Beer, composition of various kinds of, 126; origin of the word, 2; ropiness of, 115; souring of, 115; substances contained in, 113
Beers, bottled, 116, 122; carbonated, 119; chilled and filtered, 121; various classes, 6

Cane-sugar, 19
Caramel, 22
Carbohydrate, meaning of the term, 21
Carbon-dioxide, utilization of, 122
Coppers, wort, 60
Cytase, 12

Diastase, 8, 13, 15, 28

Enzymes, 12, 15; of yeast, 104, 107

Fermentation, 19, 73, 92, 108; by-products of, 103; chemical nature of, 93, 112; Liebig's and Pasteur's views on, 93; present views as to the nature of, 97; 'primary' and 'secondary', 113; systems of, 109; without yeast cells, 98
Fermenting-tun, 108, 109
Finings, 117

Galactose, 105
Glucose, 20
Grains, spent, 42

Hexose, 101
Hexose phosphate, 101
Hexosephosphatase, 101
Hop-back, 60
Hops, bitter acids, 52; early use of, in brewing, 46; essential oil of, 48; preservative properties of, 51, 56; quantities used in brewing, 61; resins, 50; storage of, 51; structure of, 46; tannin, 53
Hydrolysis, 31

Invertase, 15, 19
Invert sugar, 20

Lactase, 105
Lactose, 105
Lager beer, 57

INDEX

Lupulin, 47

Maize, flaked, 18, 38
Malt, 9
Malt, black, 22; brown, 22; grinding of, 26; modification, 28
Malting, changes occurring during, 10–15
Maltase, 105
Maltose, 30, 32, 33
Malto-dextrins, 31, 35
Mashing, 23; decoction system of, 40; infusion system of, 40
Mash-tun, 27

Original gravity, 5, 41

Peptase, 14
Porter, origin of, 5
Proteolysis, 14

Refrigerators, 65
Rice, flaked, 18, 38

Saccharomyces, anomalus, 84, 87; apiculatus, 84, 88; cerevisiae, 83; ellipsoideus II, 84, 85; foetidus, 84; ilicis, 84; Pastorianus, 114; Pastorianus I, 84, 86; Pastorianus III, 84
Sparging, 42

Starch, conversion of, by diastase, 31; soluble, 30, 31

Torulae, 118

Water, treatment of, for brewing, 25
Wort, aeration of, 66, 73; cooler, 63; cooling of the, 62; modes of infection with wild yeasts, 85; objects of boiling the, 45; protection of, from infection with undesirable organisms, 68; proteins of, 58
Wort receiver, 44

Yeast ash, composition of, 91; cell, description of, 74; enzymes of, 104, 107; from single cell, 81; nucleus of, 79; nutrition of, 89; proximate composition of, 92; pure cultures used in brewing, 87; reproduction of by ascospore formation, 76; reproduction of by budding, 76; 'top' and 'bottom,' 83; utilization of waste, 123
Yeasts, cultivated, 83; wild, 71, 83; secondary, 114

Zymase, 92; coenzyme of, 100
Zygosaccharomyces, 79

For EU product safety concerns, contact us at Calle de José Abascal, 56–1°,
28003 Madrid, Spain or eugpsr@cambridge.org.

www.ingramcontent.com/pod-product-compliance
Ingram Content Group UK Ltd.
Pitfield, Milton Keynes, MK11 3LW, UK
UKHW040157230326
469255UK00012B/145